CDT
in context

David Williamson · Tricia Sharpe

CONTENTS

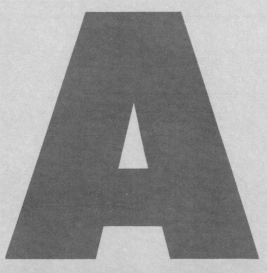

LEVEL A

INTRODUCTION

The aim of this book is to help you to develop your own confidence and skills in **craft, design and technology**. It will encourage you to look carefully at situations and problems and advise you how to investigate them. Questions and drawings suggest how you can develop your ideas. There is guidance on the planning, making and testing of your solutions.

The pages themselves show you that design sheets can be interesting, lively and easy to understand.

INTRODUCTION TO LEVEL A

The work in Level A will help you to start off well in **craft, design and technology**. Through a series of simple projects you will build up an understanding of many aspects of CDT. Each page gives you important information, so you must study both words and drawings carefully.

In some projects you will use materials like wood, metal or plastic; some use paper or card; others introduce you to simple technology. For all the projects you will need to investigate first and then work out your own ideas on paper. Your teacher will give you valuable advice. When your design is complete you should decide how best to test it so that you can report on whether it is successful.

PENCIL BOX

A large number of pens, pencils, sharpeners and other equipment used by most school children can often get untidy and disorganised. Design and make a decorative box which will hold this equipment using the materials provided.

INVESTIGATION

What sizes are the items you use?

Thickness

Th

W

LENGTH

L

WIDTH

W

SHOULD YOU MEASURE THE LARGEST OR THE SMALLEST OF THESE ITEMS?

RECORDING YOUR FINDINGS

ITEM	L	W	T
PENS			
PENCILS			
SHARPENER			
ERASER			

WHAT CAN YOU BUY?

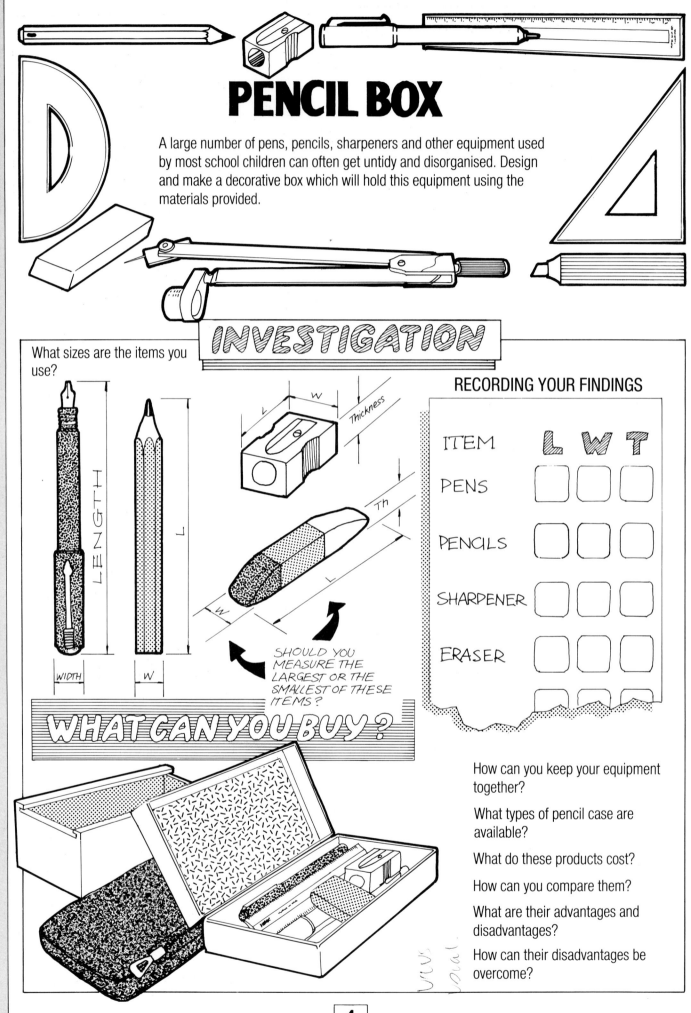

How can you keep your equipment together?

What types of pencil case are available?

What do these products cost?

How can you compare them?

What are their advantages and disadvantages?

How can their disadvantages be overcome?

PLANNING AND MAKING

How will your top be opened?

How many parts will you need?

What will the size of each part be?

Should the box have divisions inside?

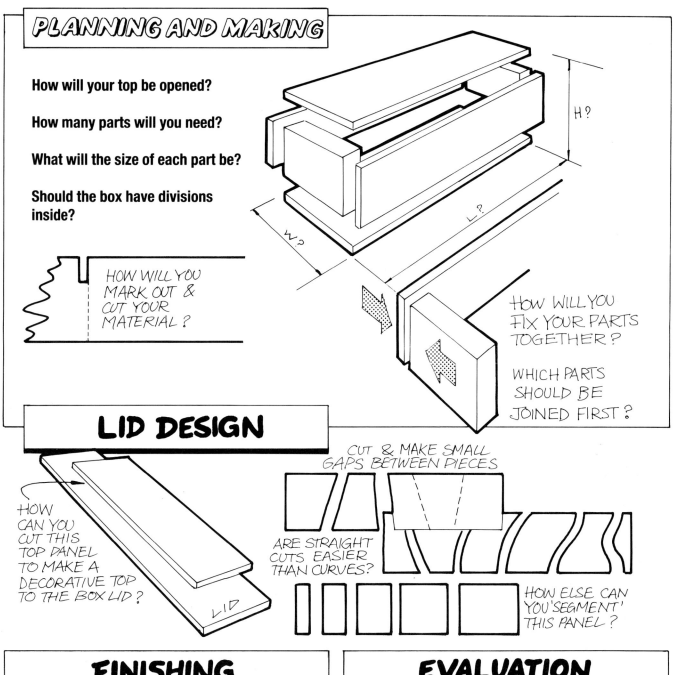

HOW WILL YOU MARK OUT & CUT YOUR MATERIAL ?

H?

L?

W?

HOW WILL YOU FIX YOUR PARTS TOGETHER ?

WHICH PARTS SHOULD BE JOINED FIRST ?

LID DESIGN

HOW CAN YOU CUT THIS TOP PANEL TO MAKE A DECORATIVE TOP TO THE BOX LID ?

LID

CUT & MAKE SMALL GAPS BETWEEN PIECES

ARE STRAIGHT CUTS EASIER THAN CURVES?

HOW ELSE CAN YOU 'SEGMENT' THIS PANEL ?

FINISHING

HOW CAN YOU TRIM THE EDGES THAT STICK OUT ?

HOW CAN YOU GET A SMOOTH FINISH ?

HOW COULD YOU APPLY COLOUR TO THE SEGMENTS?

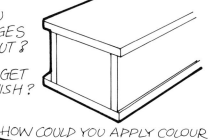

HOW CAN YOU STOP THE BOX GETTING DIRTY ?

EVALUATION

Is the box the correct size for the equipment you use?

Are you satisfied with your results?

Are there things you could improve?

How does your product compare with ones that can be bought?

How does your product compare with others made?

SITUATION

Many objects can mark or damage tables or worksurfaces in the home.

Examine a situation in your own home where this may be a problem.

DESIGN BRIEF

For the problem area you have found, design and make a mat using plywood as your material and pyrography for decoration.

INVESTIGATION

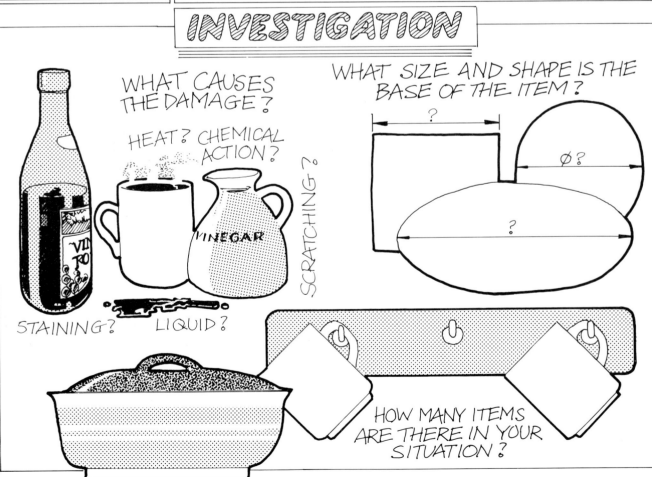

WHAT CAUSES THE DAMAGE?

HEAT? CHEMICAL ACTION?

SCRATCHING?

STAINING?   LIQUID?

VINEGAR

WHAT SIZE AND SHAPE IS THE BASE OF THE ITEM?

Ø?

HOW MANY ITEMS ARE THERE IN YOUR SITUATION?

Developing Ideas

What should you think about when deciding upon a shape for your mat?

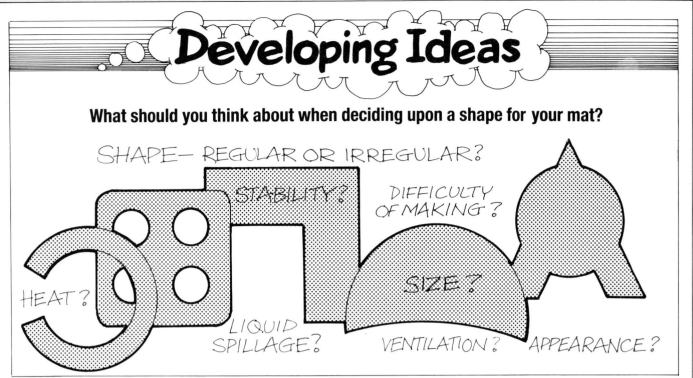

SHAPE— REGULAR OR IRREGULAR?

STABILITY?

DIFFICULTY OF MAKING?

HEAT?

SIZE?

LIQUID SPILLAGE?

VENTILATION?

APPEARANCE?

· PYROGRAPHY ·

USING HEAT TO BURN PATTERNS INTO WOOD

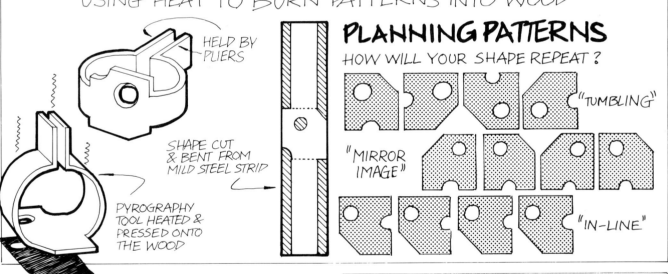

HELD BY PLIERS

SHAPE CUT & BENT FROM MILD STEEL STRIP

PYROGRAPHY TOOL HEATED & PRESSED ONTO THE WOOD

PLANNING PATTERNS

HOW WILL YOUR SHAPE REPEAT?

"TUMBLING"

"MIRROR IMAGE"

"IN-LINE"

WHAT SHAPES CAN YOU MAKE?

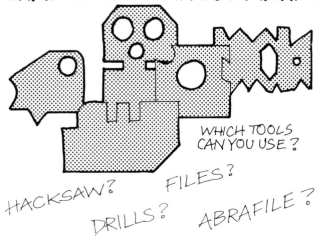

WHICH TOOLS CAN YOU USE?

HACKSAW?   FILES?

DRILLS?   ABRAFILE?

FINISHING

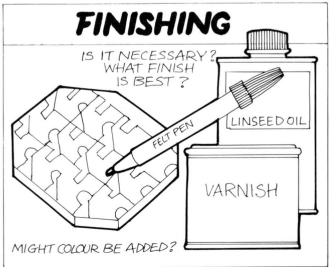

IS IT NECESSARY? WHAT FINISH IS BEST?

FELT PEN

LINSEED OIL

VARNISH

MIGHT COLOUR BE ADDED?

Most of us have many keys and each one has a different use. But it is often difficult to find the correct one quickly because they look so much alike.
Design and make an identity tag for your key(s) using one of the techniques shown.

RESEARCH

WHAT TYPES OF KEYS ARE THERE ?

WHAT DO I USE MY KEYS FOR ?

WHAT SIZES ?

HOW DO I STORE THEM ?

RECORDING

NUMBER OF TIMES USED PER WEEK

NUMBER OF TIMES

KEY NUMBER

KEY CHART

KEY SIZE AND SHAPE	USE
①	2
	30
②	

WHERE ARE KEYS KEPT ?

ON A KEYRING IN POCKET
6

20

1 UNDER A PLANT POT

5 IN A DISH IN THE KITCHEN

3 LOOSE IN A DRAWER

ANALYSIS

LOOKING AT THE PROBLEM MORE CLOSELY

WHICH KEY SHOULD I CHOOSE?

THE ONE I USE THE MOST?

THE ONE HARDEST TO IDENTIFY?

WHAT MATERIALS ARE AVAILABLE?

VENEER

ACRYLIC

PLYWOOD

COPPER

GILDING METAL

COLOUR·TEXTURE·PATTERN·SHAPE

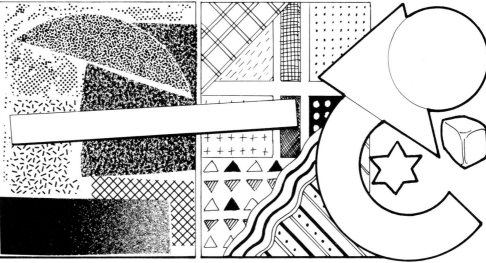

HOW CAN I FASTEN KEY TO TAG?

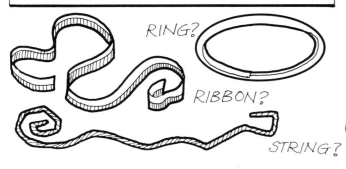

RING?

RIBBON?

STRING?

WILL STRENGTH BE IMPORTANT?

HOW CAN I STORE MY KEY AND TAG?

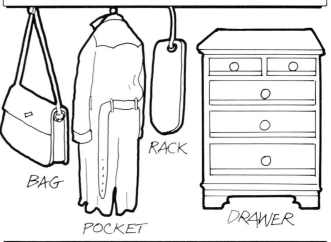

BAG

POCKET

RACK

DRAWER

SPECIFICATION
What decisions have you made?

WHAT SIZE WILL BE NEEDED?
WILL STORAGE AFFECT SIZE?

PLASTIC MEMORY

LOOKING FOR IDEAS

What does your key open?

What size tag will be needed?

ACRYLIC SHAPES

WIRE SHAPES

MODELLING

TESTING IDEAS

WILL A HOLE BE NEEDED?

WHERE?

WHICH WAY WILL THE TAG HANG?

IS THE SHAPE LARGE ENOUGH?

WHERE WILL THE INSIDE SHAPES JOIN?

USING CARD AND STRING

MAKING

ACRYLIC

1 MARK

2 CUT / FILE / WET AND DRY

3 MAKE MILD STEEL SHAPE

BDMS ROD

MEASURE CUT BEND BRAZE

FIX TO ACRYLIC

4 HEAT

5

SANDWICH

6 WET AND DRY

7 REHEAT

8 DRILL / POLISH

9

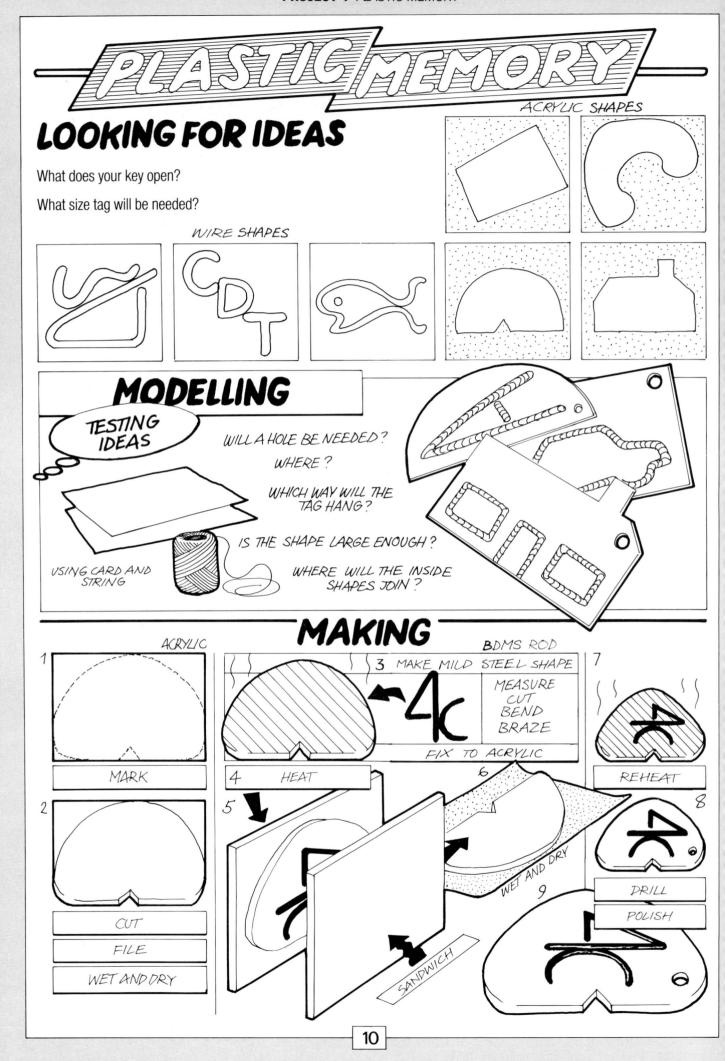

Plywood & VENEERS

LOOKING FOR IDEAS

'FOOD'

'NATURE'

'MANMADE'

MODELLING
USING CARD

GREEN CARD

RED CARD

BLACK CARD

BLACK PATTERNS AND LINES ON COLOURED CARD

Will a hole be needed?
Where will it be?
Which way will the tag hang?

MAKING
USING PLYWOOD

DOWEL 'STALK'

PLYWOOD SHAPED, COLOURED AND GLUED

ENAMEL

POWDER PAINT

COLOUR?

PYROGRAPHY

POTATO 'STAMP'

COLOURED VENEERS TO BE STUCK ON

AIR BRUSH

CAR SPRAY

VARNISH

FINISH?

COLOUR ON SPONGE DABBED ON

'MASK' (TAPE OR PLASTIC FILM REMOVED AFTER COLOURING)

11

ETCHING

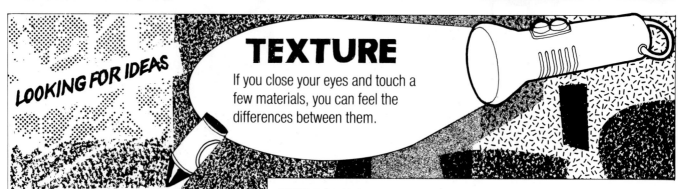

TEXTURE

If you close your eyes and touch a few materials, you can feel the differences between them.

LOOKING FOR IDEAS

COLLECTING

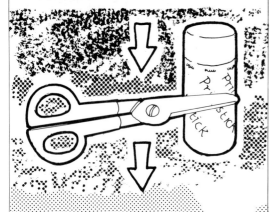

PRODUCING A RESOURCE SHEET

SELECTING

ETCHING

. . . is a way of producing a textured design on metal. Acid is used to 'eat away' all surfaces of the metal that are not protected.

MAKING

MARK OUT

SPIRIT MARKER
SCRIBER

CUT

TIN SNIPS
BENCH SHEARS

PIERCING SAW
ABRA FILE

FILE

FILES

NEEDLE FILES

CLEAN

EMERY PAPER

PAINT ON STOP OUT

SCRATCH ON DESIGN

- CLEAN OFF STOP OUT
- CENTRE PUNCH AND DRILL IF REQUIRED
- WET AND DRY PAPER CROCUS PAPER POLISH

YOUR TEACHER WILL PLACE YOUR METAL IN AN *ACID BATH*

● ENAMELLING ●

GEOMETRIC

NATURAL

COLOUR PATTERN

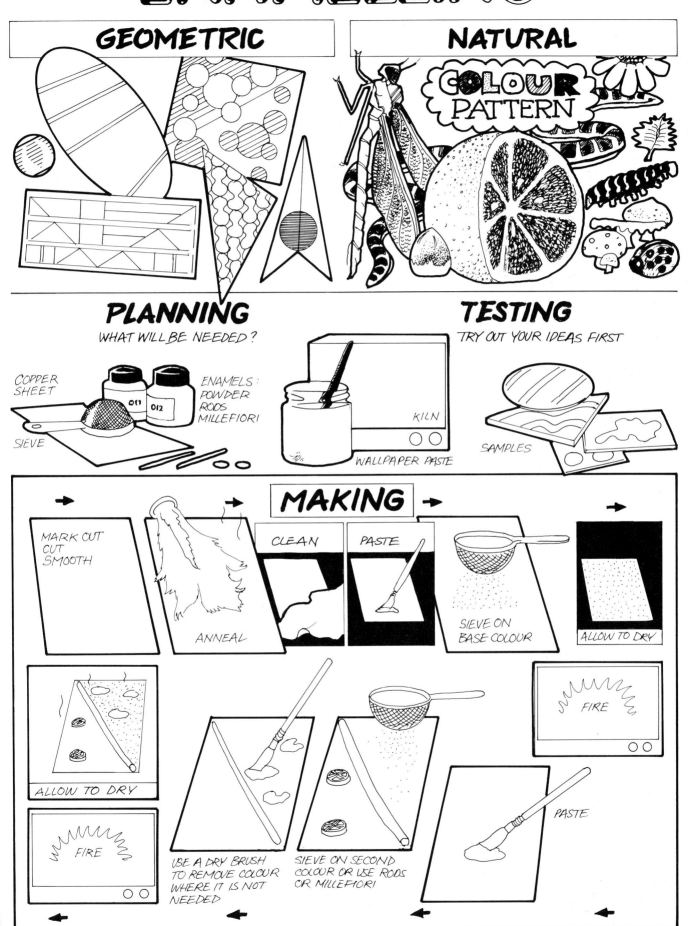

PLANNING

WHAT WILL BE NEEDED?

COPPER SHEET

ENAMELS:
POWDER
RODS
MILLEFIORI

SIEVE

011 012

KILN

WALLPAPER PASTE

TESTING

TRY OUT YOUR IDEAS FIRST

SAMPLES

MAKING

MARK OUT
CUT
SMOOTH

ANNEAL

CLEAN

PASTE

SIEVE ON
BASE COLOUR

ALLOW TO DRY

ALLOW TO DRY

FIRE

USE A DRY BRUSH
TO REMOVE COLOUR
WHERE IT IS NOT
NEEDED

SIEVE ON SECOND
COLOUR OR USE RODS
OR MILLEFIORI

FIRE

PASTE

13

STORAGE

Storing items of different sizes can be awkward. Identify a situation where storage is a problem and after investigating such things as sizes, quantity etc., design a solution using acrylic as your material.

INVESTIGATION

HOW MANY ITEMS ARE THERE?

HOW MANY COMPONENTS WILL BE NEEDED?

WHAT SIZE ARE THE ITEMS TO BE STORED?

DESIGN IDEAS

HOW CAN YOU STORE THE ITEMS?

HOW WILL THE ITEMS BE ARRANGED?

COULD HOLES BE USED?

OR PEGS?

OR MAGNETS?

OR VELCRO?

HOW CAN THE ACRYLIC BE SHAPED AND JOINED?

BENDING?

LAMINATING?

SHAPING?

MODELLING

Modelling is a very important part of any project as it quickly helps you to see your idea in 3D.

SHAPING & FORMING

What techniques will you use to shape the acrylic?

FINISHING

Acrylic can be unattractive and dangerous if edges are not finished well.

DRAWFILING?

TENSOL CEMENT

WET OR DRY PAPER?

POLISHING MACHINE?

ACRILIC POLISH?

EVALUATION

How does the finished article compare with your original idea?

What changes might you make to your design if you were starting again?

How well does it function?

What problems did you have in making your design?

What did you learn?

DESIGN BRIEF

Using softwood, design a decorative holder that will keep tidy a number of pens and pencils of different sizes.

SITUATION

LOOSE PENS AND PENCILS CAN EASILY CLUTTER A WORKSPACE

TIMETABLE

RESEARCH

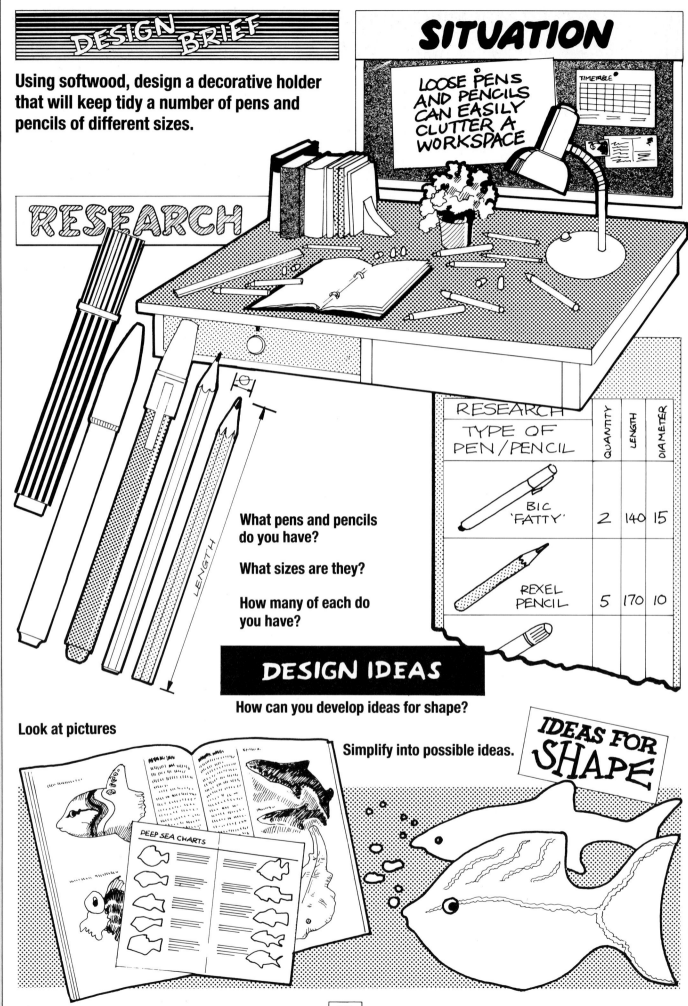

What pens and pencils do you have?

What sizes are they?

How many of each do you have?

RESEARCH TYPE OF PEN / PENCIL	QUANTITY	LENGTH	DIAMETER
BIC 'FATTY'	2	140	15
REXEL PENCIL	5	170	10

LENGTH

DESIGN IDEAS

How can you develop ideas for shape?

Look at pictures

Simplify into possible ideas.

PEEP SEA CHARTS

IDEAS FOR SHAPE

For centuries 'logos' have been used to identify
groups or individuals. Design and make a stamp
that you can use on books, envelopes, letters etc.,
that can be recognised as your individual sign.

COAT OF ARMS

TRADE MARKS

FLAGS

WHAT OTHER EXAMPLES ARE THERE?

DESIGNING YOUR OWN LOGO

Some of the logos above are very simple designs –
why do you think that is?

When designing your own logo, keep it simple. This
will make it much easier to cut out.

Can your logo be 'contained' inside a simple shape?

ERGONOMIC HANDLES

The design of handles should be given careful
thought – identify and discuss some examples.

How does the way we use tools and machines
affect the design of their handles?

Try using clay or plasticine to find the best handle
shape for your stamp.

Can the handle be simplified so it is easier to
make?

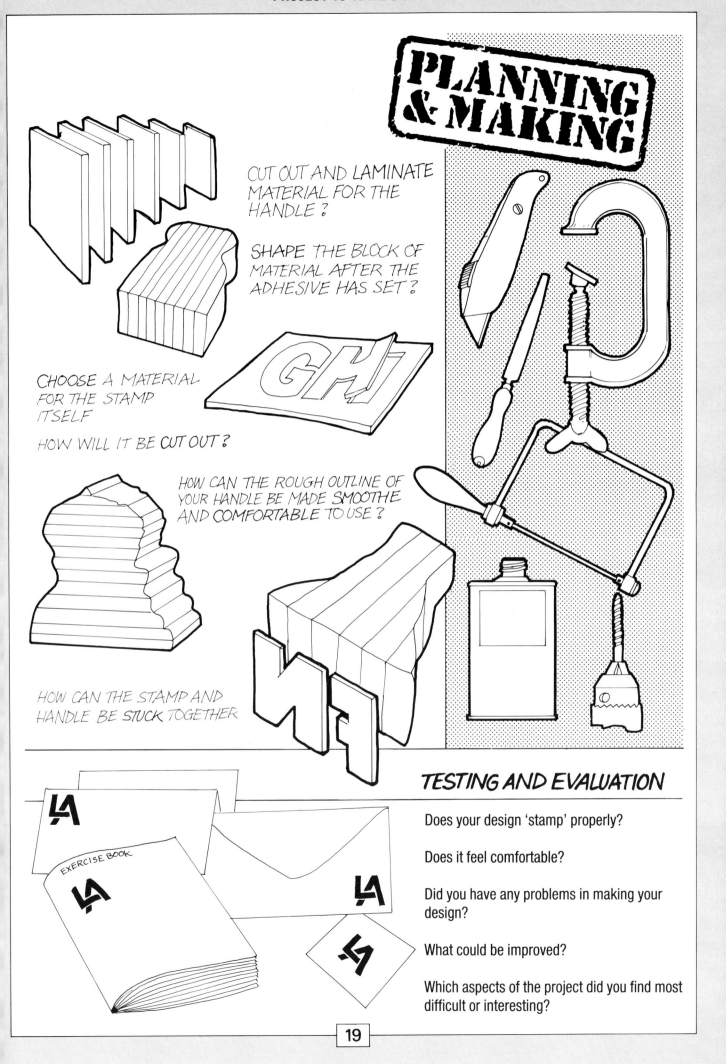

PLANNING & MAKING

CUT OUT AND LAMINATE MATERIAL FOR THE HANDLE?

SHAPE THE BLOCK OF MATERIAL AFTER THE ADHESIVE HAS SET?

CHOOSE A MATERIAL FOR THE STAMP ITSELF

HOW WILL IT BE CUT OUT?

HOW CAN THE ROUGH OUTLINE OF YOUR HANDLE BE MADE SMOOTHE AND COMFORTABLE TO USE?

HOW CAN THE STAMP AND HANDLE BE STUCK TOGETHER

EXERCISE BOOK

TESTING AND EVALUATION

Does your design 'stamp' properly?

Does it feel comfortable?

Did you have any problems in making your design?

What could be improved?

Which aspects of the project did you find most difficult or interesting?

19

Most people like to have some time to be alone.

Design and make a 'doorbell' that can be fitted to your own bedroom or another suitable place.

INVESTIGATION

WHAT CIRCUIT WILL BE USED?

CIRCUIT

MEMBRANE SWITCH

TO MAKE IT EASIER TO TURN THE CIRCUIT [ON] AND [OFF] A SWITCH IS NEEDED

CONNECT THIS SWITCH BETWEEN ONE PAIR OF WIRES

PRESS

TINFOIL

BATTERY

BUZZER

WHAT HAPPENS WHEN YOU CONNECT BLACK TO BLACK AND RED TO RED WIRES TOGETHER?

WHAT HAPPENS?
HOW DOES IT WORK?

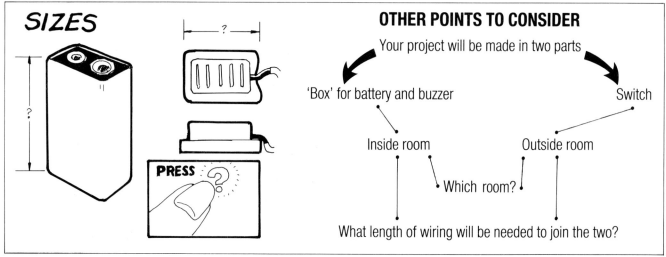

SIZES

?

?

?

PRESS

OTHER POINTS TO CONSIDER

Your project will be made in two parts

'Box' for battery and buzzer   Switch

Inside room   Outside room

Which room?

What length of wiring will be needed to join the two?

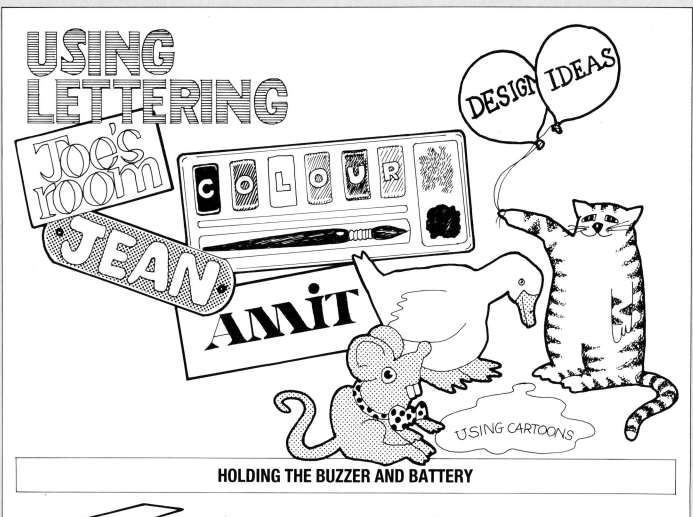

USING LETTERING

DESIGN IDEAS

Joe's room

JEAN

COLOUR

AMIT

USING CARTOONS

HOLDING THE BUZZER AND BATTERY

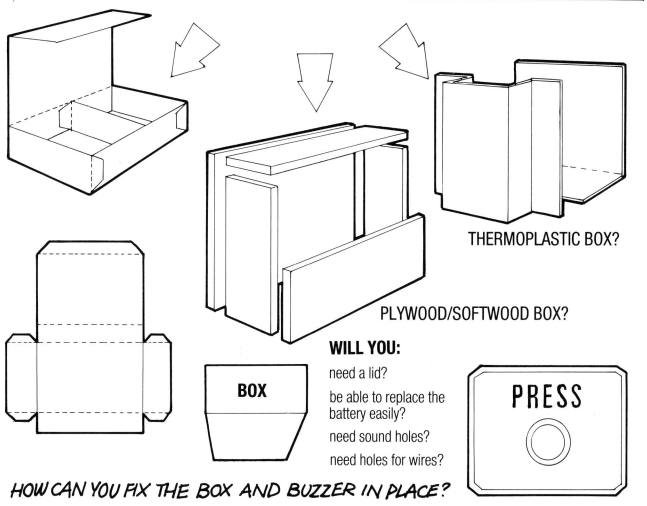

THERMOPLASTIC BOX?

PLYWOOD/SOFTWOOD BOX?

BOX

WILL YOU:

need a lid?

be able to replace the battery easily?

need sound holes?

need holes for wires?

PRESS

HOW CAN YOU FIX THE BOX AND BUZZER IN PLACE?

FLIGHT

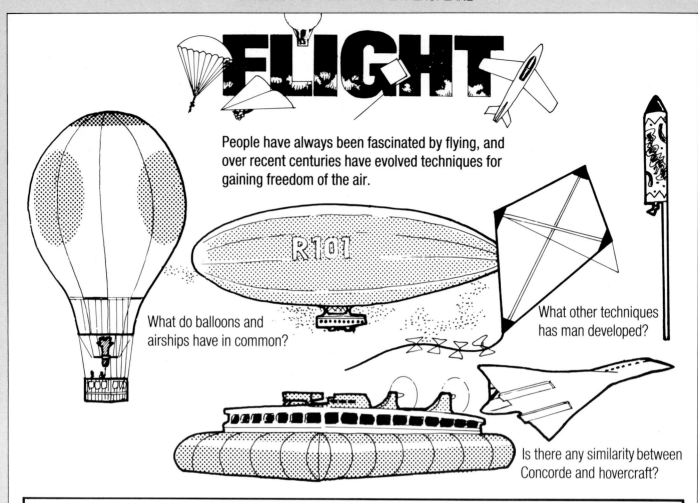

People have always been fascinated by flying, and over recent centuries have evolved techniques for gaining freedom of the air.

What do balloons and airships have in common?

What other techniques has man developed?

Is there any similarity between Concorde and hovercraft?

PROBLEM 1   PAPER AEROPLANE

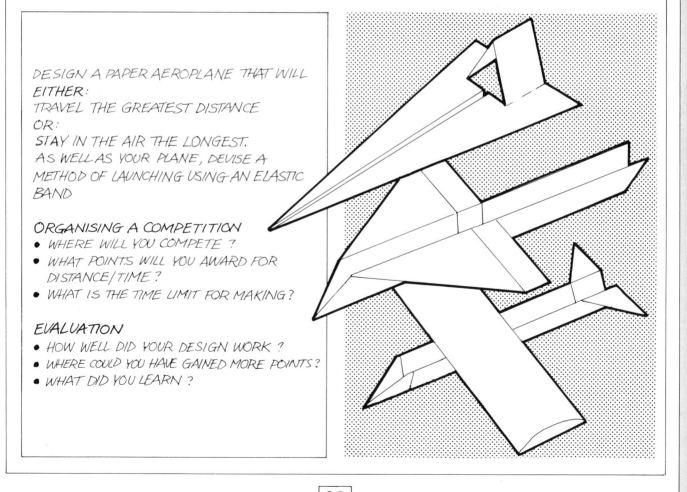

DESIGN A PAPER AEROPLANE THAT WILL
EITHER:
TRAVEL THE GREATEST DISTANCE
OR:
STAY IN THE AIR THE LONGEST.
AS WELL AS YOUR PLANE, DEVISE A
METHOD OF LAUNCHING USING AN ELASTIC
BAND

ORGANISING A COMPETITION
- WHERE WILL YOU COMPETE ?
- WHAT POINTS WILL YOU AWARD FOR DISTANCE/TIME ?
- WHAT IS THE TIME LIMIT FOR MAKING ?

EVALUATION
- HOW WELL DID YOUR DESIGN WORK ?
- WHERE COULD YOU HAVE GAINED MORE POINTS?
- WHAT DID YOU LEARN ?

PROBLEM 2 – WATER ROCKET

USING A STANDARD SIZE WASHING-UP LIQUID BOTTLE, DESIGN A 'ROCKET' THAT WILL TRAVEL THE GREATEST DISTANCE FROM A LAUNCH PAD

RECORDING AND DELIVERING
- HOW MUCH WATER PRODUCED THE BEST RESULTS?
- WHAT EFFECT DID THE ANGLE OF LAUNCH HAVE?
- HOW DID WIND DIRECTION AFFECT RESULTS?
- DID 'STREAMLINING' PRODUCE BETTER RESULTS?
- WHAT WAS SO GOOD ABOUT THE 'WINNER'?

PROBLEM 3 – KITE

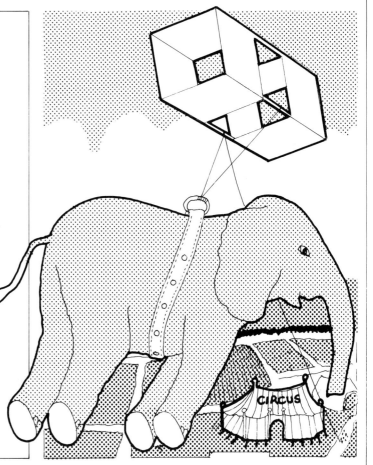

DESIGN AND MAKE A KITE THAT WILL CARRY THE HEAVIEST LOAD THE GREATEST DISTANCE DOWNWIND

EVALUATION
- WHAT KIND OF KITE PRODUCED THE BEST RESULTS?
- WHY DID IT PRODUCE THE BEST RESULTS?
- WHAT WAS THE HEAVIEST WEIGHT YOUR KITE LIFTED AND HOW COULD IT BE INCREASED?
- WHAT WERE THE BEST MATERIALS FOR MAKING THE KITE?
- HOW DID WIND SPEED AFFECT THE EFFICIENCY OF THE KITE?

THE GREAT
POLO MINT
COMPETITION

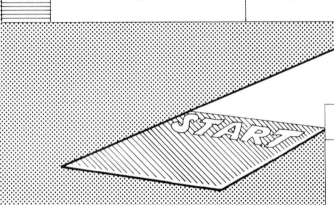

| | 10 | | 20 | 30 | 50 100 | 20 |

OBJECT OF THE GAME

USING THE MATERIALS PROVIDED PROJECT A POLO MINT ALONG A SET COURSE TO A TARGET 1·5 METERS AWAY.
THE WINNER WILL BE THE ONE WHO GAINS THE HIGHEST MARKS FROM 2 ATTEMPTS.

RULES AND REGULATIONS

★ Competitors must build their firing device on a standard block 80 × 80 mm.

★ Polo mints must remain unbroken during competition rounds.

★ The only means of propulsion should be a standard elastic band.

★ The device should not stick out from the start area.

MATERIALS

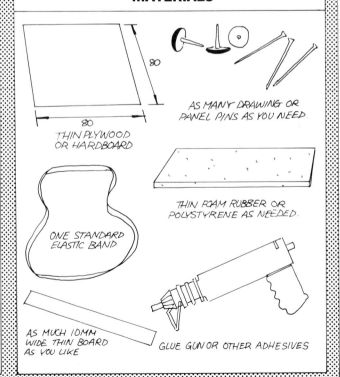

80

80
THIN PLYWOOD OR HARDBOARD

AS MANY DRAWING OR PANEL PINS AS YOU NEED.

THIN FOAM RUBBER OR POLYSTYRENE AS NEEDED.

ONE STANDARD ELASTIC BAND

AS MUCH 10MM WIDE THIN BOARD AS YOU LIKE

GLUE GUN OR OTHER ADHESIVES

·THE DAILY BLURB·

OCTOBER 7

THE PRINTED WORD CORPORATION NEWS LTD

SHOCK! TRAGEDY!

THOUSANDS LOSE HOMES IN TATTLESVILLE DISASTER

Tragedy struck yesterday when an earthquake hit Tattlesville, newspaper capital of the world. Even after considerable warning of seismic disturbance, it was still a miracle that no one was killed when virtually all of the city's buildings tumbled to the ground. All that remains of the once bustling city is a pile of rubble and thousands of undelivered newspapers.

Authorities are concerned now about the sheltering of the homeless thousands who face the onslaught of searing heat today as temperatures climb to a record high.

CALL FOR ACTION

Clarke Kent of our Metropolis office is reporting on the situation in Tattlesville and has called for additional help from structural designers.

Although communications were bad when Clarke reported, we understand that operations are being co-ordinated by a Mr Sooberman, who has requested assistance in the construction of temporary shelters for the local population. All ideas for this vast task will be gratefully accepted.

GOVERNMENT OUTCRY AT EEC POLICIES

Brussels has once more been the target for criticism. Euro M.P. R. Geldorf lashed out at policies which subsidise manufacturers of office supplies, creating yet more stockpiles.

The 'stationery mountain' is costing the European Community millions of pounds each year to store and maintain.

Adhesive tape, string, paper clips and staples are lying idle in huge warehouses throughout Europe with no obvious means of reducing stocks in the short term.

ELASTIC POWER

Investigate the ways in which the energy stored in an elastic band can be used to propel a model vehicle along a flat surface for the maximum distance.

Before starting to make the model, decide as a group on the number of points will be awarded for qualities such as speed, distance, economy of materials, appearance and whatever else you might consider important.

DESIGN CONSIDERATIONS

BODY CONSTRUCTION

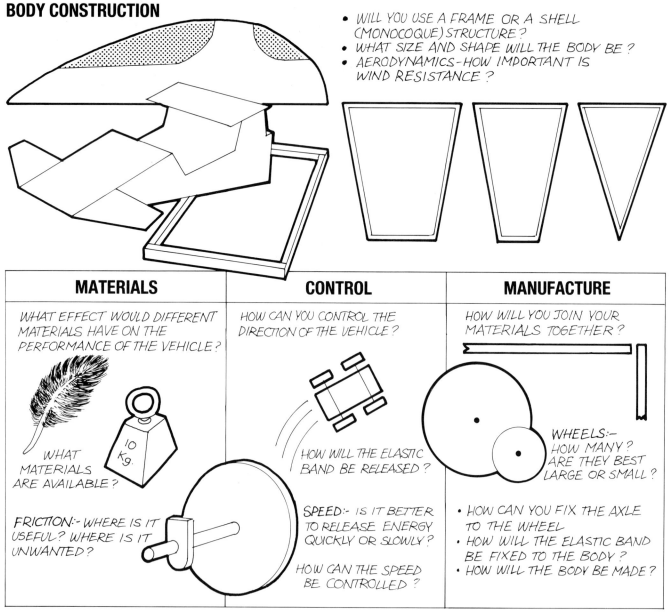

- WILL YOU USE A FRAME OR A SHELL (MONOCOQUE) STRUCTURE?
- WHAT SIZE AND SHAPE WILL THE BODY BE?
- AERODYNAMICS - HOW IMPORTANT IS WIND RESISTANCE?

MATERIALS

WHAT EFFECT WOULD DIFFERENT MATERIALS HAVE ON THE PERFORMANCE OF THE VEHICLE?

WHAT MATERIALS ARE AVAILABLE?

10 Kg.

FRICTION:- WHERE IS IT USEFUL? WHERE IS IT UNWANTED?

CONTROL

HOW CAN YOU CONTROL THE DIRECTION OF THE VEHICLE?

HOW WILL THE ELASTIC BAND BE RELEASED?

SPEED:- IS IT BETTER TO RELEASE ENERGY QUICKLY OR SLOWLY?

HOW CAN THE SPEED BE CONTROLLED?

MANUFACTURE

HOW WILL YOU JOIN YOUR MATERIALS TOGETHER?

WHEELS:- HOW MANY? ARE THEY BEST LARGE OR SMALL?

- HOW CAN YOU FIX THE AXLE TO THE WHEEL
- HOW WILL THE ELASTIC BAND BE FIXED TO THE BODY?
- HOW WILL THE BODY BE MADE?

EVALUATION

★ What distance was travelled?

★ What time did it take?

★ What was the speed of the vehicle?

★ Was it best to release the energy quickly or slowly?

★ What factors affected the performance of the vehicle?

★ What would you change in your design if you were to start again?

LEVEL

B

INTRODUCTION TO LEVEL B

Level B expects you to look at the design situation more closely than in Level A, and to analyse in more detail. The results of this should enable you to develop a wider range of solutions which extend your first ideas.

With your Level A experience, you should feel more confident in designing, and you can be more adventurous in your style of presentation. By now you will have used several materials and will know how to shape, join and finish them; this will help you to choose what is most suitable for making each design. More new materials and processes will be used at Level B.

You will have realised how important **evaluation** is and your discoveries from earlier projects should help you to avoid past mistakes.

NOTEPAD

Taking messages on the telephone often involves a scramble for paper and pen. How can this problem be overcome?

WHAT DO YOU NEED TO FIND OUT ABOUT?

RESEARCH HELPS YOU TO UNDERSTAND THE PROBLEM BETTER. ASK YOURSELF AS MANY QUESTIONS AS YOU CAN: EG —

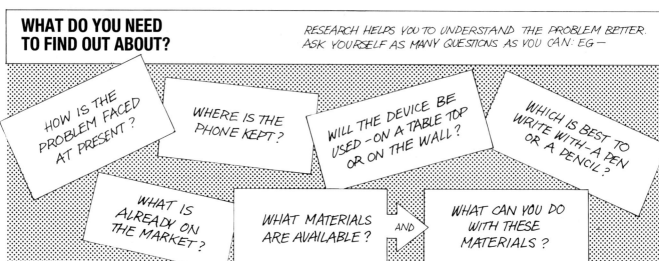

HOW IS THE PROBLEM FACED AT PRESENT?

WHERE IS THE PHONE KEPT?

WILL THE DEVICE BE USED – ON A TABLE TOP OR ON THE WALL?

WHICH IS BEST TO WRITE WITH – A PEN OR A PENCIL?

WHAT IS ALREADY ON THE MARKET?

WHAT MATERIALS ARE AVAILABLE? AND WHAT CAN YOU DO WITH THESE MATERIALS?

RECORDING INFORMATION

IT IS IMPORTANT TO KEEP A RECORD OF YOUR RESEARCH IN ORDER TO REFER TO THE INFORMATION WHEN DESIGNING. THERE ARE MANY WAYS IT CAN BE DONE. HERE ARE SOME EXAMPLES.

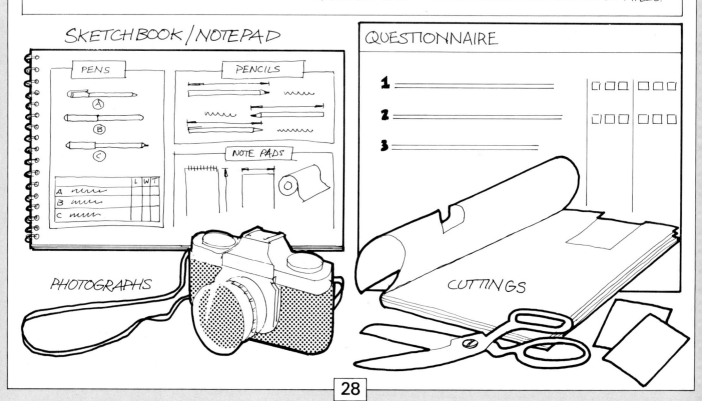

SKETCHBOOK / NOTEPAD

QUESTIONNAIRE

PHOTOGRAPHS

CUTTINGS

A N A L Y S I S

— BREAKING THE PROBLEM DOWN INTO SMALLER PARTS.

NOTEPAD OR TILL ROLL?

What sizes are available and which would be best?

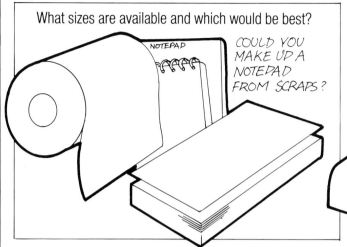

COULD YOU MAKE UP A NOTEPAD FROM SCRAPS?

SHAPE

What shape might make your design more interesting?

How can it be achieved?

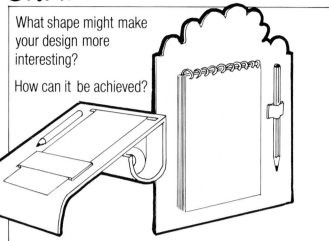

ATTACHING PEN/PENCIL

HOW CAN YOU STOP IT GETTING 'LOST'?

BENDING AND GROOVING

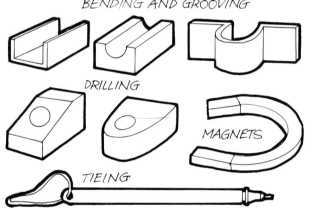

DRILLING

MAGNETS

TIEING

ATTACHING NOTEPAD

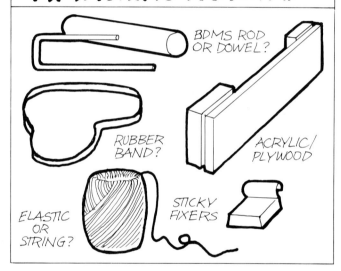

BDMS ROD OR DOWEL?

RUBBER BAND?

ACRYLIC/ PLYWOOD

ELASTIC OR STRING?

STICKY FIXERS

JOINING

HOW WILL ALL THE BITS GO TOGETHER?

RIVETS *SCREWS*

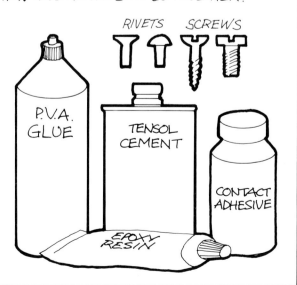

P.V.A. GLUE

TENSOL CEMENT

CONTACT ADHESIVE

EPOXY RESIN

FINISHING

ENAMEL PAINTS? *FELT PEN? WOOD STAIN?*

PYROGRAPHY?

CAR SPRAY? *VENEER?* *PLASTIC MEMORY?*

EVALUATION

Do you think your solution is successful?

How does it compare with other solutions?

What improvements could be made to your design?

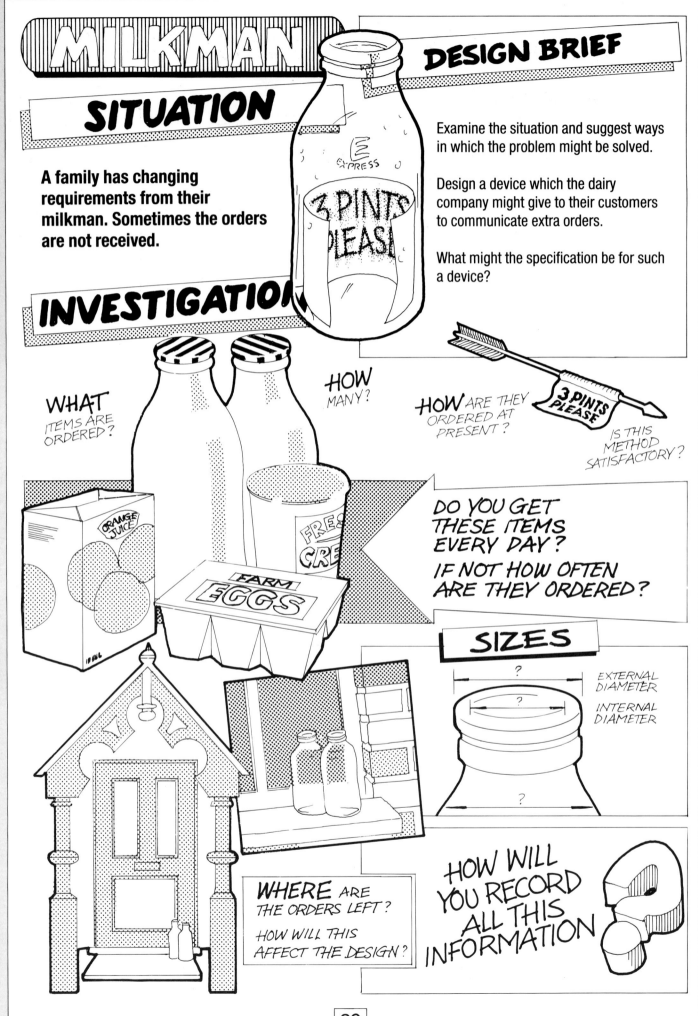

MILKMAN

SITUATION

A family has changing requirements from their milkman. Sometimes the orders are not received.

INVESTIGATION

DESIGN BRIEF

Examine the situation and suggest ways in which the problem might be solved.

Design a device which the dairy company might give to their customers to communicate extra orders.

What might the specification be for such a device?

3 PINTS PLEASE

WHAT ITEMS ARE ORDERED?

HOW MANY?

HOW ARE THEY ORDERED AT PRESENT?

3 PINTS PLEASE

IS THIS METHOD SATISFACTORY?

ORANGE JUICE

FRESH CREAM

FARM EGGS

DO YOU GET THESE ITEMS EVERY DAY?

IF NOT HOW OFTEN ARE THEY ORDERED?

SIZES

?
EXTERNAL DIAMETER

?
INTERNAL DIAMETER

?

WHERE ARE THE ORDERS LEFT?

HOW WILL THIS AFFECT THE DESIGN?

HOW WILL YOU RECORD ALL THIS INFORMATION

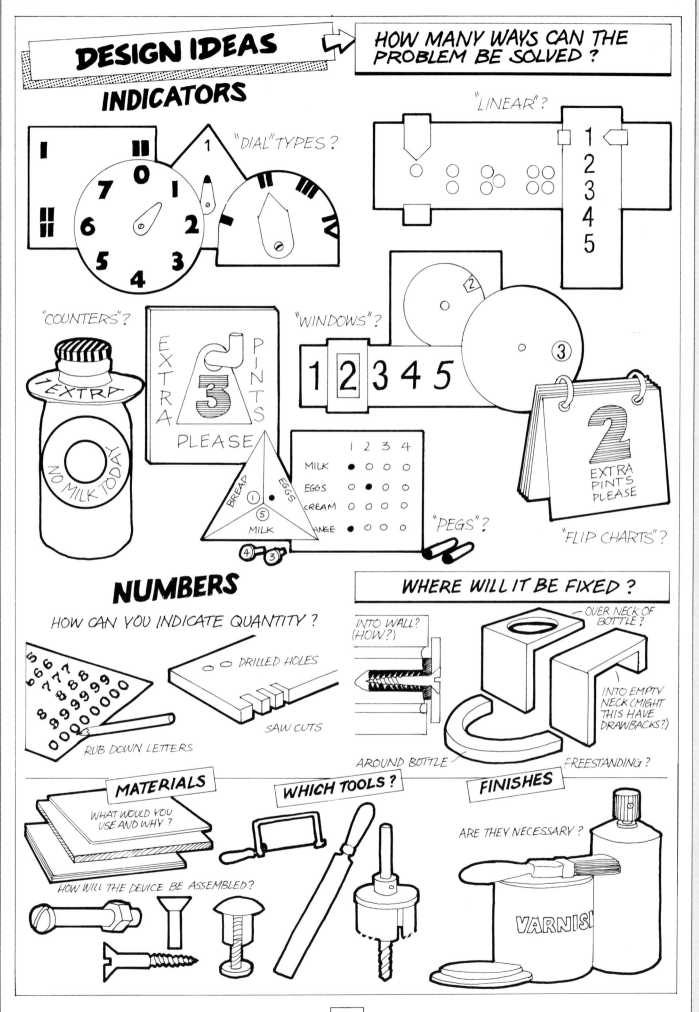

DESIGN IDEAS

HOW MANY WAYS CAN THE PROBLEM BE SOLVED?

INDICATORS

"DIAL" TYPES?

"LINEAR"?

"COUNTERS"?

"WINDOWS"?

EXTRA PINTS PLEASE

1 EXTRA
NO MILK TODAY

	1	2	3	4
MILK	●	○	○	○
EGGS	○	●	○	○
CREAM	○	○	○	○
...NGE	●	○	○	○

BREAD EGGS MILK

"PEGS"?

2 EXTRA PINTS PLEASE

"FLIP CHARTS"?

NUMBERS

HOW CAN YOU INDICATE QUANTITY?

RUB DOWN LETTERS

○ ○ DRILLED HOLES

SAW CUTS

WHERE WILL IT BE FIXED?

INTO WALL? (HOW?)

— OVER NECK OF BOTTLE?

INTO EMPTY NECK (MIGHT THIS HAVE DRAWBACKS?)

AROUND BOTTLE

FREESTANDING?

MATERIALS

WHAT WOULD YOU USE AND WHY?

HOW WILL THE DEVICE BE ASSEMBLED?

WHICH TOOLS?

FINISHES

ARE THEY NECESSARY?

VARNISH

MIRROR
MIRROR
ON THE
WALL...

People need to use mirrors for many different purposes. Investigate where and why mirrors are used, and from the results of your research identify a situation for which you can design a solution.

INVESTIGATION

| THINK ABOUT WHEN WE USE MIRRORS | ASK OTHER PEOPLE | KEEP OUR EYES OPEN FOR IDEAS |

RECORDING INFORMATION

ANALYSIS

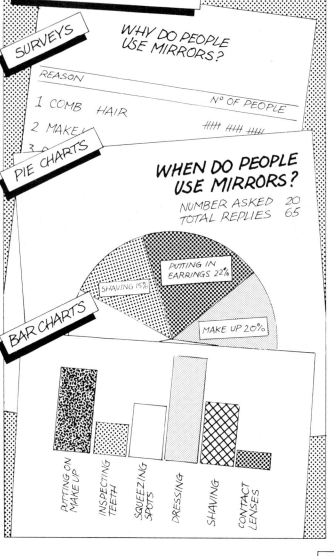

SURVEYS

WHY DO PEOPLE USE MIRRORS?

REASON   N° OF PEOPLE

1 COMB HAIR
2 MAKE   ### ### ###

PIE CHARTS

WHEN DO PEOPLE USE MIRRORS?

NUMBER ASKED   20
TOTAL REPLIES 65

PUTTING IN EARRINGS 22%
SHAVING 15%
MAKE UP 20%

BAR CHARTS

PUTTING ON MAKE UP · INSPECTING TEETH · SQUEEZING SPOTS · DRESSING · SHAVING · CONTACT LENSES

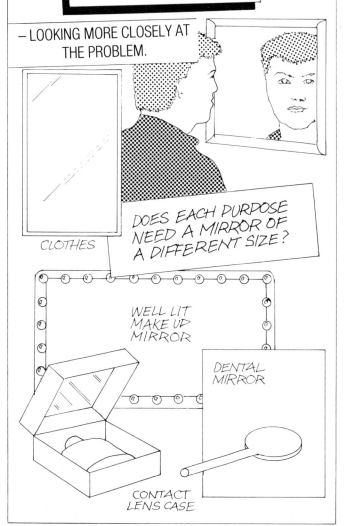

– LOOKING MORE CLOSELY AT THE PROBLEM.

CLOTHES

DOES EACH PURPOSE NEED A MIRROR OF A DIFFERENT SIZE?

WELL LIT MAKE UP MIRROR

DENTAL MIRROR

CONTACT LENS CASE

WHERE WILL IT BE USED?

FIXED? → TO WALL? → HANGING?
→ TO DOOR? → PERMANENT?

FREESTANDING? → ON SHELF?
→ ON DRESSING TABLE?

H?W

HOW WILL THE MIRROR BE FRAMED?

DESIGN IDEAS

STRAIGHT CUTS AND DRILLED HOLES

PLYWOOD CUT-OUTS

SHAPED MIRROR AND COMPLICATED CUT-OUT

APPLIED SHAPES OVER BASIC FRAME

HOW CAN THE MIRROR BE HELD IN THE FRAME ?

MIRROR SANDWICHED BETWEEN PLYWOOD

SMALL CLIPS

CORNER PLATES

WHAT OTHER METHODS ARE THERE?

PLANNING AND MANUFACTURE

CUTTING → FITTING

SMOOTHING ← JOINING

DECORATING COLOURING → FINISHING

EVALUATING

33

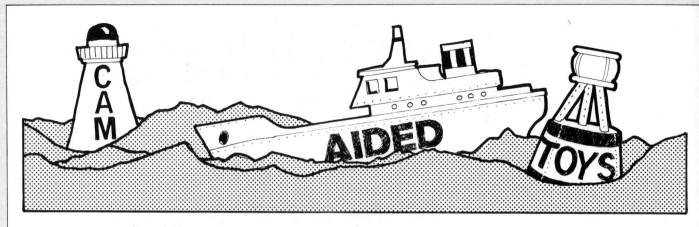

Both adults and children are fascinated by movement in toys. Sometimes these have been developed into shop-window displays which are used to catch the eye. Research subjects suitable for animation using cams and make a cam-aided 'toy'.

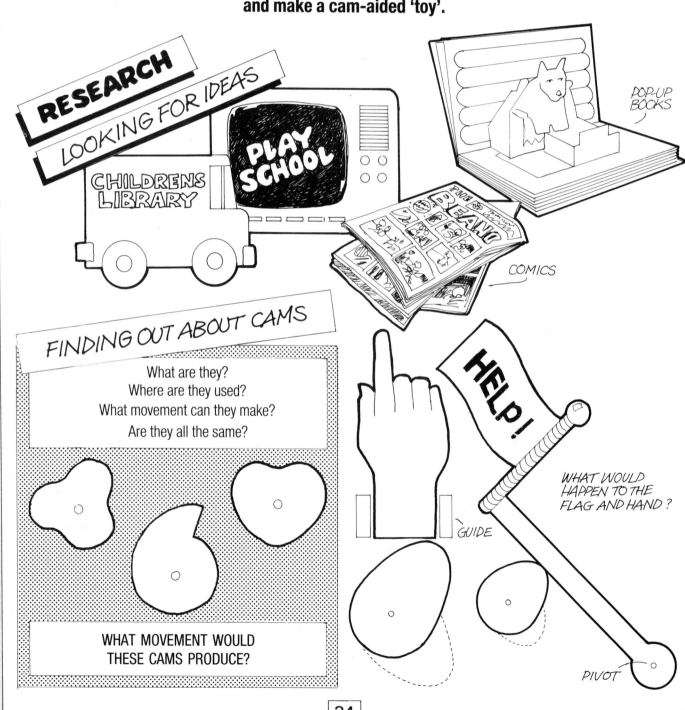

RESEARCH

LOOKING FOR IDEAS

CHILDRENS LIBRARY

PLAY SCHOOL

POP-UP BOOKS

COMICS

FINDING OUT ABOUT CAMS

What are they?
Where are they used?
What movement can they make?
Are they all the same?

WHAT MOVEMENT WOULD THESE CAMS PRODUCE?

HELP!

GUIDE

WHAT WOULD HAPPEN TO THE FLAG AND HAND?

PIVOT

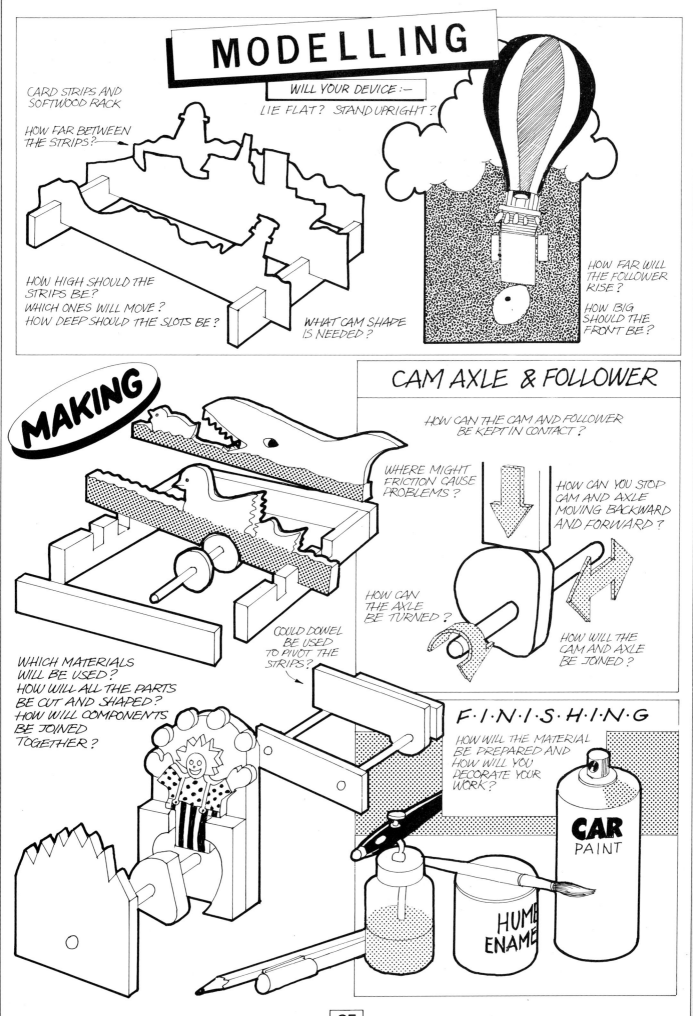

MODELLING

WILL YOUR DEVICE :—
LIE FLAT? STAND UPRIGHT?

CARD STRIPS AND SOFTWOOD RACK

HOW FAR BETWEEN THE STRIPS?

HOW HIGH SHOULD THE STRIPS BE?
WHICH ONES WILL MOVE?
HOW DEEP SHOULD THE SLOTS BE?

WHAT CAM SHAPE IS NEEDED?

HOW FAR WILL THE FOLLOWER RISE?

HOW BIG SHOULD THE FRONT BE?

MAKING

WHICH MATERIALS WILL BE USED?
HOW WILL ALL THE PARTS BE CUT AND SHAPED?
HOW WILL COMPONENTS BE JOINED TOGETHER?

COULD DOWEL BE USED TO PIVOT THE STRIPS?

CAM AXLE & FOLLOWER

HOW CAN THE CAM AND FOLLOWER BE KEPT IN CONTACT?

WHERE MIGHT FRICTION CAUSE PROBLEMS?

HOW CAN YOU STOP CAM AND AXLE MOVING BACKWARD AND FORWARD?

HOW CAN THE AXLE BE TURNED?

HOW WILL THE CAM AND AXLE BE JOINED?

F·I·N·I·S·H·I·N·G

HOW WILL THE MATERIAL BE PREPARED AND HOW WILL YOU DECORATE YOUR WORK?

CAR PAINT

HUMB ENAME

Children like picking up and fitting pieces together. This activity helps them learn how to 'manipulate' objects.
Design and make a simple plywood jigsaw for a young child that also helps them learn a second skill.

RESEARCH

HOW DO YOUNG CHILDREN PLAY ?
HOW DO YOUNG CHILDREN LEARN ?

WATCH A YOUNG CHILD

WHAT COLOURS DO THEY LIKE ?

HOW MANY PIECES CAN THEY COPE WITH BEFORE THEY LOSE INTEREST ?

WHAT SHAPES DO THEY LIKE PICKING UP ?

WHAT SIZE PIECES CAN THEY HOLD ?

WHAT KIND OF HELP DO THEY NEED TO PUT JIGSAWS TOGETHER ?

WHAT SIZE ARE THEIR HANDS ?

WHAT ARE THEY LEARNING ?

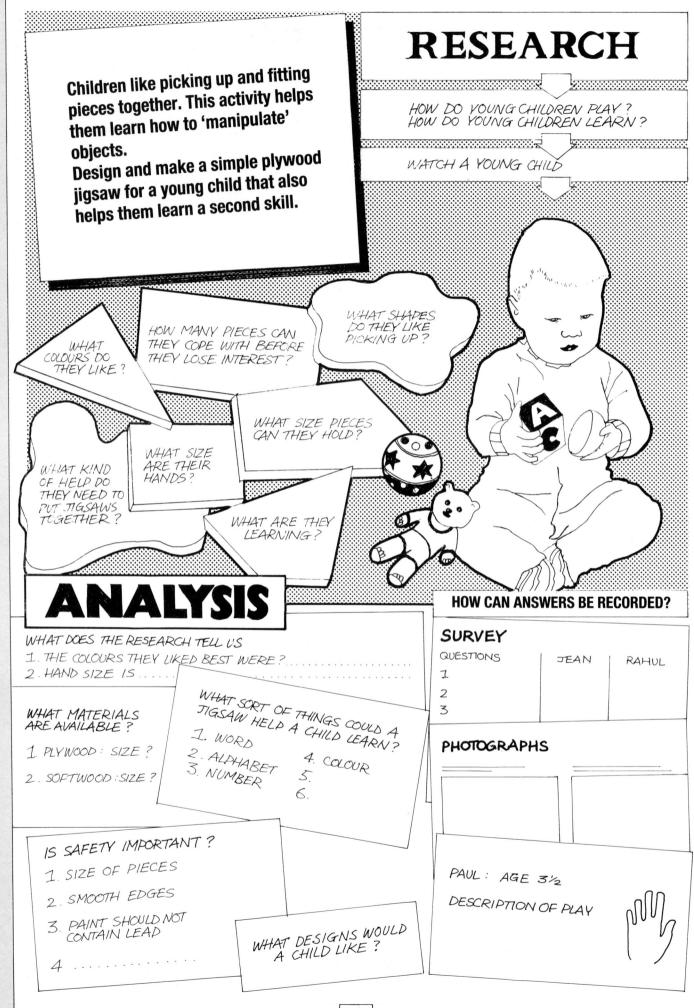

ANALYSIS

WHAT DOES THE RESEARCH TELL US
1. THE COLOURS THEY LIKED BEST WERE ? . . . .
2. HAND SIZE IS . . . .

WHAT MATERIALS ARE AVAILABLE ?

1. PLYWOOD : SIZE ?

2. SOFTWOOD : SIZE ?

WHAT SORT OF THINGS COULD A JIGSAW HELP A CHILD LEARN ?
1. WORD
2. ALPHABET   4. COLOUR
3. NUMBER   5.
                 6.

IS SAFETY IMPORTANT ?

1. SIZE OF PIECES

2. SMOOTH EDGES

3. PAINT SHOULD NOT CONTAIN LEAD

4 . . . .

WHAT DESIGNS WOULD A CHILD LIKE ?

HOW CAN ANSWERS BE RECORDED?

SURVEY

QUESTIONS	JEAN	RAHUL
1		
2		
3		

PHOTOGRAPHS

PAUL : AGE 3½

DESCRIPTION OF PLAY

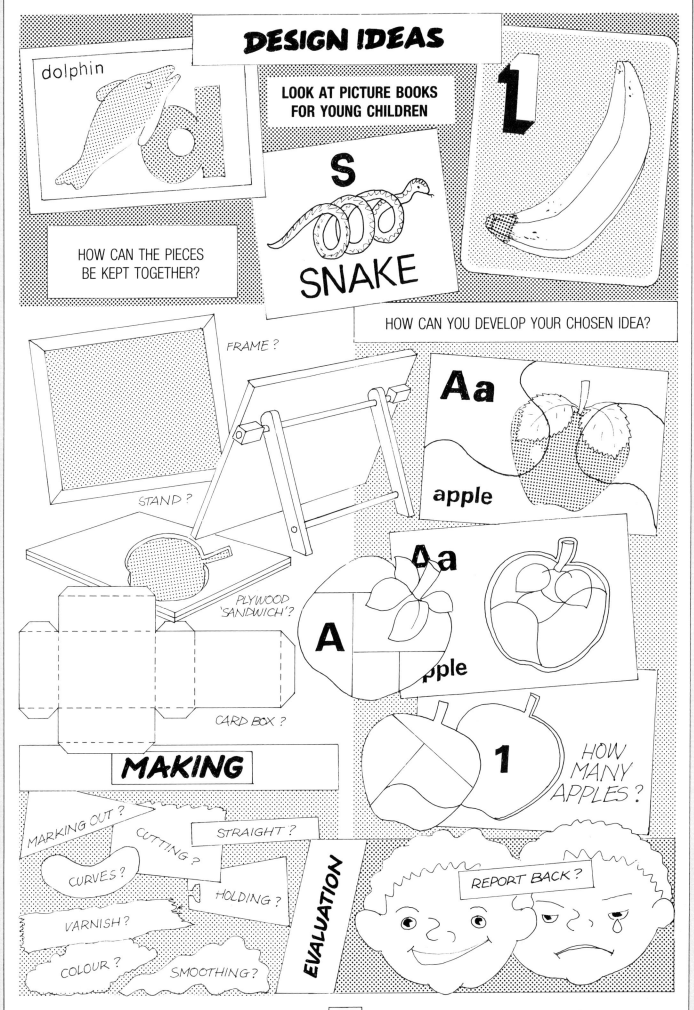

DESIGN IDEAS

dolphin

LOOK AT PICTURE BOOKS FOR YOUNG CHILDREN

HOW CAN THE PIECES BE KEPT TOGETHER?

S
SNAKE

HOW CAN YOU DEVELOP YOUR CHOSEN IDEA?

FRAME ?

STAND ?

PLYWOOD 'SANDWICH'?

CARD BOX ?

Aa
apple

Aa
apple

A

1 HOW MANY APPLES ?

MAKING

MARKING OUT ?

CUTTING ?

STRAIGHT ?

CURVES ?

HOLDING ?

VARNISH ?

EVALUATION

COLOUR ?

SMOOTHING ?

REPORT BACK ?

MONEY BOX

Most people find it difficult to save money. Using the mechanism given, design and make a softwood moneybox that encourages people to put money into it.

INVESTIGATE

THE MECHANISM GIVEN

What happens when a coin is dropped into the box?

How is the movement transferred from inside the box to the front of the box?

What is the purpose of the 'guide'?

How can you work out the balance?

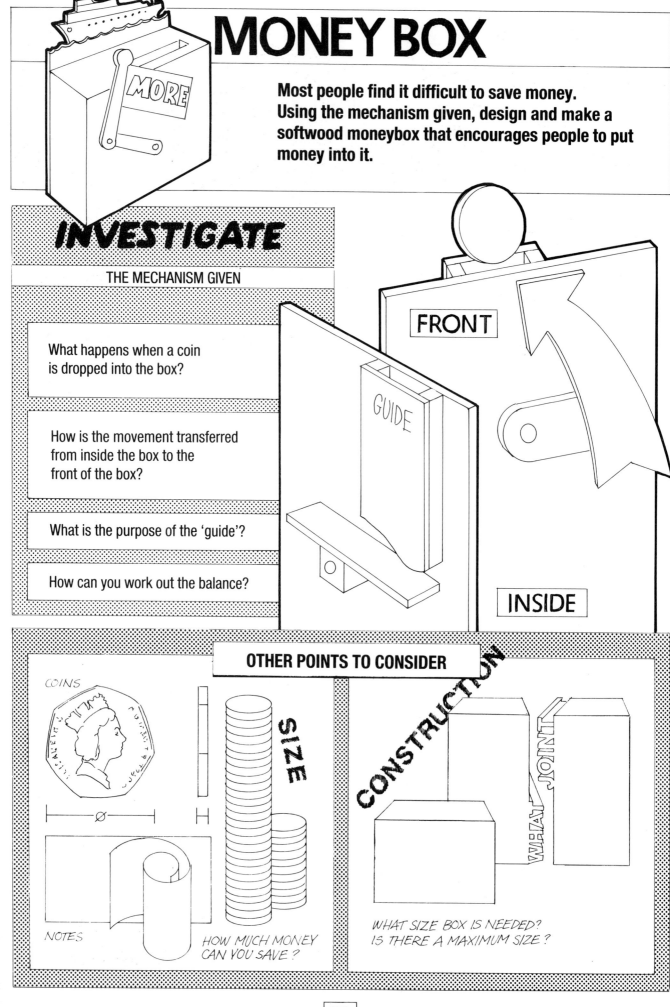

FRONT

GUIDE

INSIDE

OTHER POINTS TO CONSIDER

COINS

SIZE

NOTES

HOW MUCH MONEY CAN YOU SAVE?

CONSTRUCTION

WHAT JOINT

WHAT SIZE BOX IS NEEDED? IS THERE A MAXIMUM SIZE?

THINK ABOUT IDEAS

LOOK AT PICTURE BOOKS

POP UPS

USING IDEAS

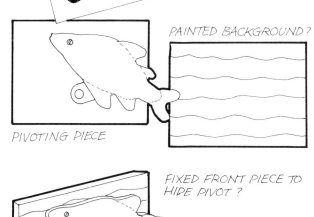

PAINTED BACKGROUND?

PIVOTING PIECE

FIXED FRONT PIECE TO HIDE PIVOT?

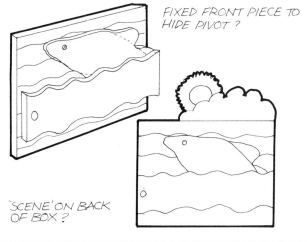

'SCENE' ON BACK OF BOX?

GETTING MONEY OUT

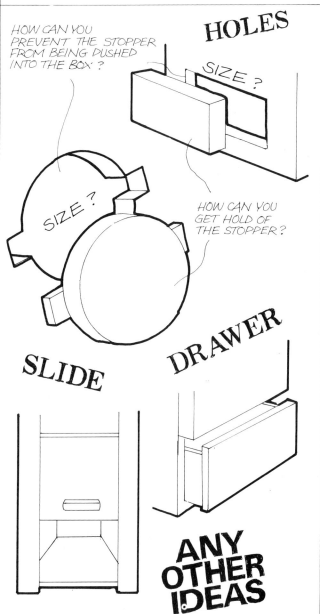

HOW CAN YOU PREVENT THE STOPPER FROM BEING PUSHED INTO THE BOX?

HOLES

SIZE?

SIZE?

HOW CAN YOU GET HOLD OF THE STOPPER?

SLIDE

DRAWER

ANY OTHER IDEAS

PLANNING

What should you make first:

Box?

Front with mechanism?

EVALUATION

Without box

With box

Money collected in one month in pence.

39

SPECTACULAR WARDROBE

Spectacles and sunglasses are part of personal 'image'.
Design and model a selection of glasses to be worn by a
stage performer who tries to put over an outrageous image.

WHY DO PEOPLE THINK ABOUT THEIR IMAGE?

OCCUPATION	LIFESTYLE	PERSONALITY

BUSINESS PERSON

CLACTON

LIFEGUARD

PUNK

SUPERSTAR

HIPPY

HAPPY

SAD

MYSTERIOUS

MAD

'ANTHROPOMETRICS' – HUMAN MEASUREMENT

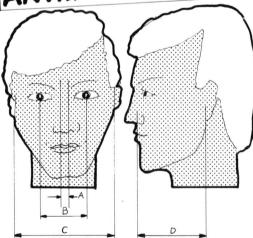

**HOW CAN YOU TAKE
MEASUREMENTS FROM
THE HUMAN HEAD?**

Ⓐ WIDTH OF NOSE
Ⓑ DISTANCE BETWEEN EYES
Ⓒ WIDTH OF HEAD
Ⓓ DISTANCE FROM NOSE TO EAR

Does everybody have the same head measurement?

How much do they differ?

What size would fit most people?

How can you show the results of your survey clearly?

STARTING POINTS

FOOD.... SPORT.... NATURE.... TRANSPORT....
CAN YOU THINK OF OTHER AREAS THAT MIGHT INSPIRE IDEAS?

HOW WILL THE SPECS BE KEPT ON?

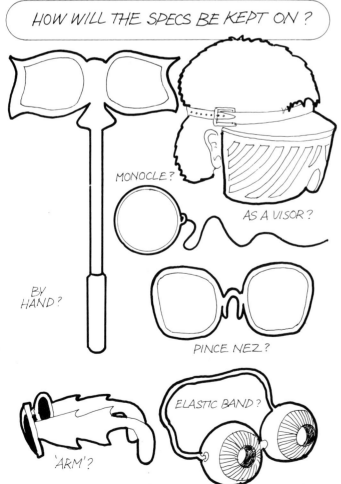

MONOCLE?

AS A VISOR?

BY HAND?

PINCE NEZ?

'ARM'?

ELASTIC BAND?

MODELLING AND MAKING

What factors might affect your choice of material?

What materials might you use?

Which is best and why?

How will you make your design?

How will you achieve a good finish?

Does safety need to be considered?

How can you judge your results?

CARDS CARDS CARDS CARDS CARDS

Many people like to give cards to their family and friends on special occasions.

All of us like to receive cards that are unusual.

Design and make an unusual card to give on a special occasion.

ANALYSIS AND RESEARCH

Who is the card for?

What design of card would they like?

What materials are available?

If it is to be posted, will this affect the design?

What occasion?

What size should it be?

What images are usually used for the occasion?

What original ideas can you think up?

HOW CAN YOU MAKE YOUR CARD UNUSUAL?

POP-UPS

A VERY GOOD PLACE TO LOOK FOR IDEAS IS IN POP-UP BOOKS

SOME CARDS USE ELASTIC BANDS TO PRODUCE A POP-UP EFFECT

TURN TO MOVE EYES

PULL TAG TO MAKE LEG MOVE

BACK

BASE

ELASTIC BANDS

A MERRY XMAS

A COMMON POP-UP METHOD

CUT-OUTS

YOU ARE 3

LETTERING

WHAT STYLE WILL BE SUITABLE?

LOOK AT...

CARDS

BOOK COVERS

POSTERS

STENCILS

DRY TRANSFERS

PLANNING AND MAKING

- WHAT MATERIALS & TOOLS WILL YOU NEED?
- WHAT SAFETY FACTORS MUST YOU CONSIDER WHEN MAKING?
- HOW WILL YOU APPLY COLOUR?
- DO YOU NEED SPECIAL GLUES?

WHAT ELSE SHOULD YOU CONSIDER?

'Image' plays an important part in the success or failure of pop stars. Hundreds of records are released each week and each one has to capture attention in order to sell. Most bands develop an image to use for record promotion, and it is communicated through record sleeves, videos etc.

Name a new band of your own and create an image for it. Title its first single and then produce both a record sleeve – 7" or 12" – to help promote it, and a stage set suitable for its performance.

RESEARCH

HOW DO BANDS CREATE AN IMAGE?

DESIGN IDEAS

WHAT SORT OF IMAGE?

CARD
'CUT OUTS'
PAINT
MARKERS

SINGLE OF THE WEEK

HOT IMPORT

OUT NOW

NEW SINGLE

MATERIALS

O U T · N O W

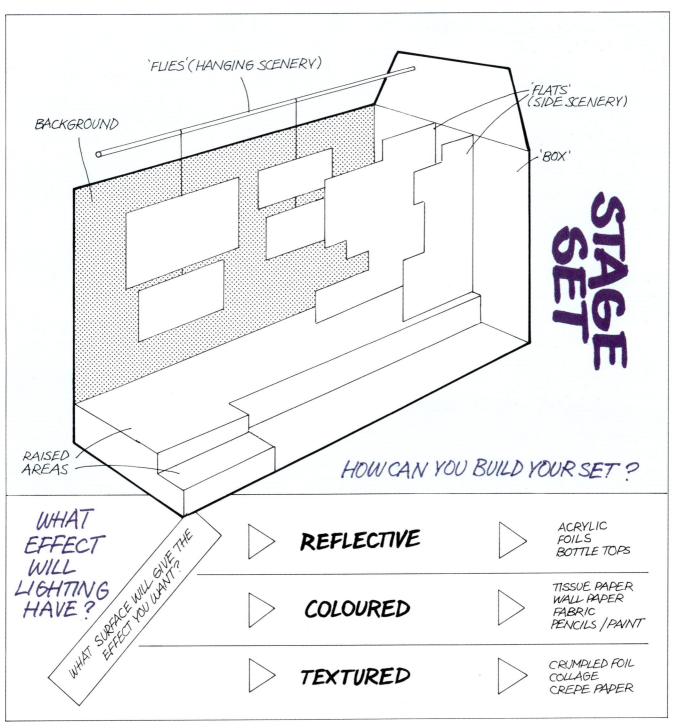

'FLIES' (HANGING SCENERY)

BACKGROUND

'FLATS'
(SIDE SCENERY)

'BOX'

STAGE SET

RAISED AREAS

HOW CAN YOU BUILD YOUR SET?

WHAT
EFFECT
WILL
LIGHTING
HAVE?

WHAT SURFACE WILL GIVE THE EFFECT YOU WANT?

▷ **REFLECTIVE** ▷ ACRYLIC
FOILS
BOTTLE TOPS

▷ **COLOURED** ▷ TISSUE PAPER
WALL PAPER
FABRIC
PENCILS / PAINT

▷ **TEXTURED** ▷ CRUMPLED FOIL
COLLAGE
CREPE PAPER

45

SITUATION

'Finishing' is an important part of every practical project, and yet it is often made more difficult because suitable abrasive tools are not available.

Design and make a hand tool to hold abrasive paper that will help you 'finish' a particular project.

INVESTIGATION ASKING QUESTIONS AND COLLECTING INFORMATION

ERGONOMICS – THE STUDY OF HUMAN BEINGS AND THEIR ENVIRONMENT.

HOW CAN THE 'BEST GRIP' BE FOUND

HOW WILL THE TOOL BE USED?

GRIPPED BY THE HAND

PUSHED DOWN ON TO WORK

PUSHED ACROSS WORK

  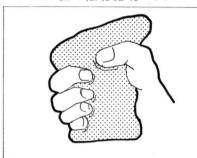

'MODELLING'

HOW CAN A PERMANENT RECORD BE MADE? ▶KEEP MODELS? ▶SKETCHES? ▶PHOTOGRAPHS?

WHEN MIGHT SUCH A TOOL BE USEFUL?

LARGE FLAT SURFACES

CURVED SURFACES

SMALL SPACES

EDGES

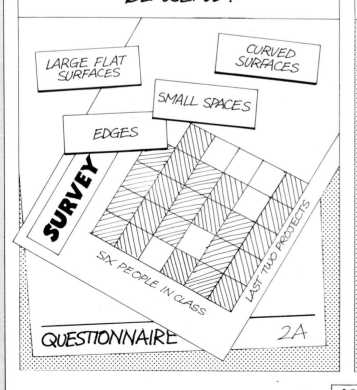

SURVEY

SIX PEOPLE IN CLASS

LAST TWO PROJECTS

QUESTIONNAIRE 2A

WHAT HAND TOOLS ARE THERE THAT DO THIS JOB ALREADY?

HAND TOOL	GOOD POINTS	BAD POINTS
CORK SANDING BLOCK		GLASS PAPER HAS TO BE HELD IN PLACE BY FINGERS. NO USE ON SOME CURVES OR IN SMALL SPACES
SPONGE ABRASIVE BLOCK		

ANALYSIS

-BREAKING THE PROBLEM DOWN INTO SMALLER PARTS.

WHAT SHAPE SHOULD THE HOLDER BE?

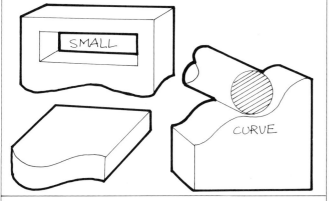

SMALL

CURVE

WHAT SURFACE WILL IT BE USED ON?

ABRASIVE PAPER

GLASS PAPER   EMERY PAPER   WET AND DRY PAPER

Which one should I use?
What sizes is it available in?
Should my holder be adaptable for all types?

RUBBER BAND?   SLIDE INTO SLOTS?

GLUED ON?

VELCRO?

SCREW ON STRIP?

FINGERS?   SELOTAPE?

Can the paper be replaced easily?
Is the paper used economically?

ERGONOMIC SHAPE FOR HANDLE.

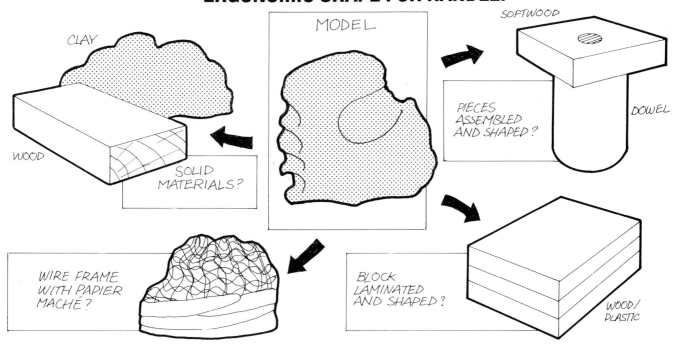

CLAY

MODEL

SOFTWOOD

WOOD

SOLID MATERIALS?

PIECES ASSEMBLED AND SHAPED?

DOWEL

WIRE FRAME WITH PAPIER MACHÉ?

BLOCK LAMINATED AND SHAPED?

WOOD/ PLASTIC

MANUFACTURE

Which tools will be needed?
In what order should pieces be made?
What finish will be most suitable?

TESTING AND EVALUATION

How well does it work?
How comfortable is it to hold and use?
Can you suggest improvements?

555 CHRISTMAS

At Christmas many people like to decorate their homes inside and out.

Design a light-up Christmas decoration using LEDs.

HOW CAN I MAKE THE LIGHTS FLASH ON AND OFF?

WHAT CIRCUIT CAN I USE?

WHAT COMPONENTS WILL BE USED?

WHAT WILL BE THE TOTAL COST?

HOW CAN THE CIRCUIT BE TURNED ON AND OFF?

CIRCUITS

1

330 R

WHAT HAPPENS WHEN YOU CLOSE THE SWITCH?

CAN YOU ADD MORE LEDS?

2 TO BUILD THIS CIRCUIT YOU WILL NEED A 555 TIMER

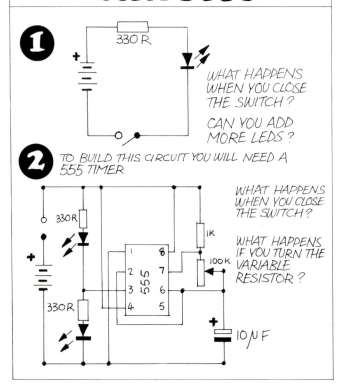

330 R

1K

100 K

555

330 R

10 µF

WHAT HAPPENS WHEN YOU CLOSE THE SWITCH?

WHAT HAPPENS IF YOU TURN THE VARIABLE RESISTOR?

SWITCHES

SWITCHES → WHEN OPEN BREAK THE CIRCUIT AND TURN IT OFF
→ WHEN CLOSED CONDUCT ELECTRICITY AND TURN THE CIRCUIT ON

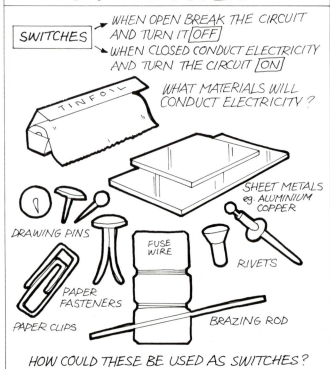

WHAT MATERIALS WILL CONDUCT ELECTRICITY?

TINFOIL

SHEET METALS eg. ALUMINIUM COPPER

DRAWING PINS

FUSE WIRE

RIVETS

PAPER FASTENERS

PAPER CLIPS

BRAZING ROD

HOW COULD THESE BE USED AS SWITCHES?

48

ANALYSIS

WHERE WILL THE DECORATION HANG OR STAND?

WILL IT BE USED INDOORS OR OUTSIDE?

WHERE CAN YOU LOOK FOR IDEAS?

HOW WILL IT HANG OR STAND?

WHAT MATERIALS ARE AVAILABLE FOR USE?

WHAT SIZE AND COLOURS ARE THE LEDS?

SIZES...

HOUSING THE BATTERY AND CIRCUIT...

SOFTWOOD

THIN PLY

ACRYLIC BENT ON STRIP HEATER

VACUUM OR PRESS-FORMED STYRENE

DEVELOPING IDEAS...

WHAT IMAGES WILL BE IMPROVED WITH FLASHING LIGHTS?

HOW WILL YOU...

★ cut your materials to shape?
★ ensure a good finish?
★ colour your masterpiece?

★ join different materials?
★ evaluate your results?

For at least three thousand years, people have found it necessary to have devices that measure time. Gradually these methods have become more sophisticated until now we can measure down to a tiny fraction of a second.

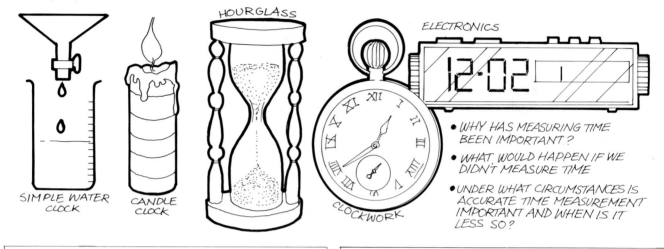

SIMPLE WATER CLOCK

CANDLE CLOCK

HOURGLASS

CLOCKWORK

ELECTRONICS

- WHY HAS MEASURING TIME BEEN IMPORTANT?
- WHAT WOULD HAPPEN IF WE DIDN'T MEASURE TIME
- UNDER WHAT CIRCUMSTANCES IS ACCURATE TIME MEASUREMENT IMPORTANT AND WHEN IS IT LESS SO?

PROBLEM 1

GOOD DESIGN IS ABOUT GETTING THE BEST RESULT AT THE LEAST COST.
FOR THE PROBLEM BELOW DEVISE A COSTING SYSTEM THAT CAN BE USED IN EVALUATING EFFECTIVENESS OF DESIGN AGAINST COST:

e.g. STRAWS   =   1 POINT EACH
     CUPS   =   5 POINTS EACH
     CARD   =   10 POINTS PER SHEET

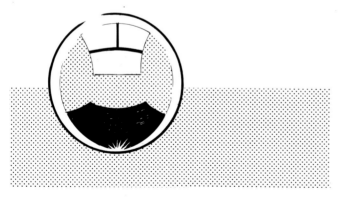

USING THE MINIMUM OF MATERIALS, DESIGN A STRUCTURE WHICH WILL ENABLE A BALL BEARING, PUT IN AT ONE POINT, TO EXIT AT ANOTHER EXACTLY 10 SECONDS LATER. THE BALL BEARING MUST BE MOVING FOR THE _WHOLE_ 10 SECONDS

PROBLEM 2

USING HEAT, AIR, LIGHT, WATER OR A COMBINATION OF ALL AS YOUR ENERGY SOURCE, DEVISE A MEANS OF SOUNDING AN ALARM AFTER EXACTLY 60 SECONDS. USE WHATEVER MATERIALS ARE AVAILABLE TO YOU.

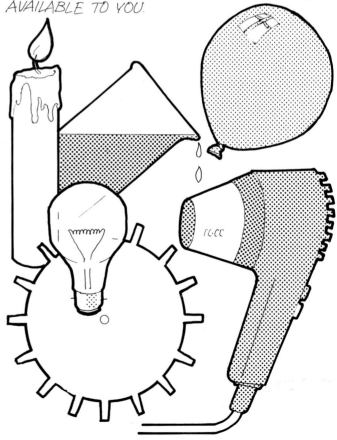

PROBLEM 3 – WATCH DESIGN

EVEN IN HISTORY, TIMEPIECES HAVE RARELY BEEN PURELY 'FUNCTIONAL' MORE OFTEN THAN NOT THEY WERE CONSIDERED EXAMPLES OF THE CRAFTSPERSON'S SKILL, AND WERE DESIGNED ACCORDINGLY.

THE WATCHES WE WEAR TODAY HAVE IN THE SAME WAY, BECOME ASSOCIATED WITH FASHION AND PERSONAL 'IMAGE'. DESIGN AND MODEL A WATCH FOR A PARTICULAR KIND OF PERSON.

ANALYSIS AND INVESTIGATION

WHO WILL YOU DESIGN FOR?

MODEL? YOUNG? EXECUTIVE?

"STYLE"

YUPPIE? COOL? TRENDY? NURSE?

ERGONOMICS
HOW CAN YOU FIND THE BEST STRAP SIZE?

WHAT IS A COMFORTABLE SIZE FOR THE FACE?

ANALOGUE OR DIGITAL

5:55 PM

WHICH IS EASIEST TO READ? WHICH SUITS THE DESIGN THE BEST?

WHAT IS ON THE MARKET ALREADY?

WHO DO THE DESIGNS APPEAL TO?

HOW DOES THE DESIGN APPEAL TO THIS GROUP?

HOW DOES THE COST SUIT THE TARGET GROUP?

WHAT IMAGES ARE USED IN ADVERTISING THE PRODUCT?

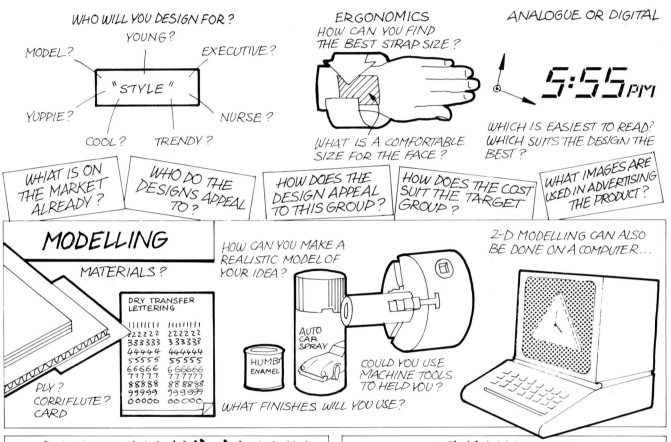

MODELLING

MATERIALS?

DRY TRANSFER LETTERING

PLY? CORRIFLUTE? CARD

HOW CAN YOU MAKE A REALISTIC MODEL OF YOUR IDEA?

AUTO CAR SPRAY

HUMBROL ENAMEL

COULD YOU USE MACHINE TOOLS TO HELP YOU?

WHAT FINISHES WILL YOU USE?

2-D MODELLING CAN ALSO BE DONE ON A COMPUTER...

PACKAGING AND DISPLAY

CAN PACKAGING AND DISPLAY BE COMBINED? WHAT INFORMATION DO YOU WANT TO PUT OVER? HOW CAN YOU SHOW YOUR PRODUCT TO BEST ADVANTAGE?

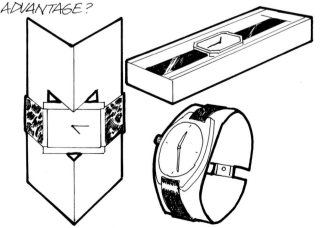

EVALUATION

How successful was your idea?

How would you improve your design further?

What 'criteria' would you use for comparing designs?

Who could evaluate the designs as 'outside' observers?

How can you compare results between different ideas?

What have you learnt from the project?

PROBLEM 4 – ELECTRONIC TIMER

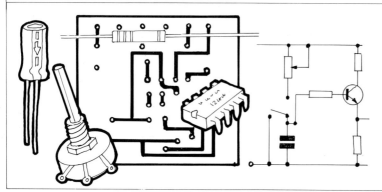

The development of electronic timers has made it relatively simple to produce accurate time delays.

Identify a situation where a time delay alarm would aid the accuracy of a process. Design both a suitable circuit and its casing.

SITUATIONS

WHERE MIGHT A TIMER BE NEEDED?
KITCHEN?
GAMES?

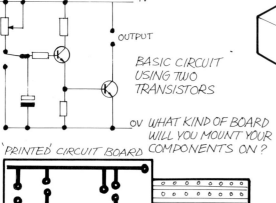

TELEPHONE?   DARKROOM?

TIME DELAY

ELECTRONIC TIMING IS PRODUCED BY USING A RESISTOR AND CAPACITOR IN SERIES.

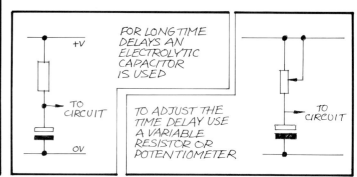

FOR LONG TIME DELAYS AN ELECTROLYTIC CAPACITOR IS USED

TO ADJUST THE TIME DELAY USE A VARIABLE RESISTOR OR POTENTIOMETER

THE CIRCUIT

THE SIMPLEST CIRCUIT USES TRANSISTORS, BUT YOU COULD USE A NUMBER OF INTEGRATED CIRCUITS (MICROCHIPS) TO DO THE SAME THING.

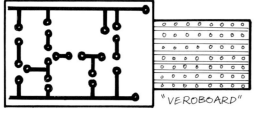

BASIC CIRCUIT USING TWO TRANSISTORS

WHAT KIND OF BOARD WILL YOU MOUNT YOUR COMPONENTS ON?

'PRINTED' CIRCUIT BOARD

"VEROBOARD"

THE CASING

WHAT WILL THE CONTAINER TO HOLD THE CIRCUIT LOOK LIKE?

HOW WILL YOU :-
FIT ALL THE PARTS IN PLACE?
REPLACE THE BATTERIES?
MAKE YOUR BOX AND A KNOB?

COULD YOU GET IDEAS FROM EXISTING PRODUCT DESIGN (RADIOS, PERSONAL STEREOS, CLOCKS ETC)?

HOW WILL IT BE 'STYLED'?

EVALUATION

★ How accurate is it?

★ How well made is it?

★ What have you learnt when making your timer?

★ How does your device compare with commercial ones?

★ Why are commercial versions cheaper to make?

★ What should be considered in the costing of your design?

There are many ways of applying colour to your design sheets. They all have advantages and disadvantages. Here are some examples that you might be able to use. Experiment to find out which is best for you.

COLOURED PENCIL IS ONE OF THE MOST COMMON METHODS OF APPLYING COLOUR TO DESIGN SHEETS. IT IS EASY TO USE AND GRADUAL CHANGES IN COLOUR AND SHADE CAN BE ACHIEVED. WHAT ARE ITS DISADVANTAGES?

FELT-TIP PENS ARE ANOTHER POPULAR MEDIUM. LOW COST AND WIDE VARIETY OF COLOURS ARE THEIR MAIN ADVANTAGES - WHAT ARE THE DRAWBACKS?

WATER COLOURS AND INKS CAN GIVE A VERY ATTRACTIVE 'WASH' ON DESIGN SHEETS. CAN THEY BE USED ON ALL PAPERS?

AIRBRUSHING CAN GIVE A VERY PROFESSIONAL FINISH TO WORK. HOW EASY IS IT TO USE?

COLOURED PAPERS, PASTEL COLOURS, CAR SPRAY AND SPIRIT-BASED MARKERS ARE OTHER METHODS OF APPLYING COLOUR. IN WHAT WAY ARE THE BEST USED?

EVERY PICTURE TELLS A STORY

Visual images are used for many different purposes, one of which is to communicate information. Look at the images on these pages and consider each of these questions:

1 What information is being conveyed?

2 Is it easily understood?

3 How does the graphic image convey information?

4 What graphic methods are used?

5 How suitable or effective are these chosen methods?

PHOTOGRAPHS

PRESENTATION DRAWINGS

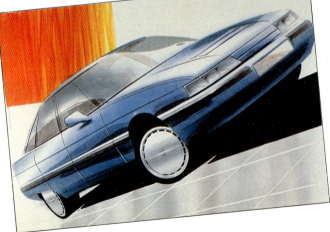

TECHNICAL DRAWINGS

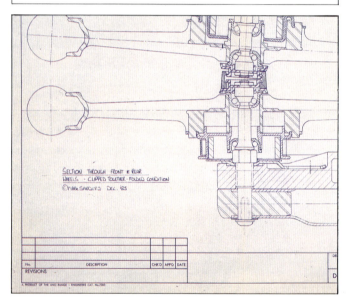

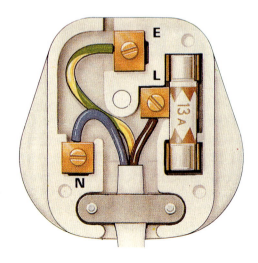

STATISTICAL REPRESENTATION

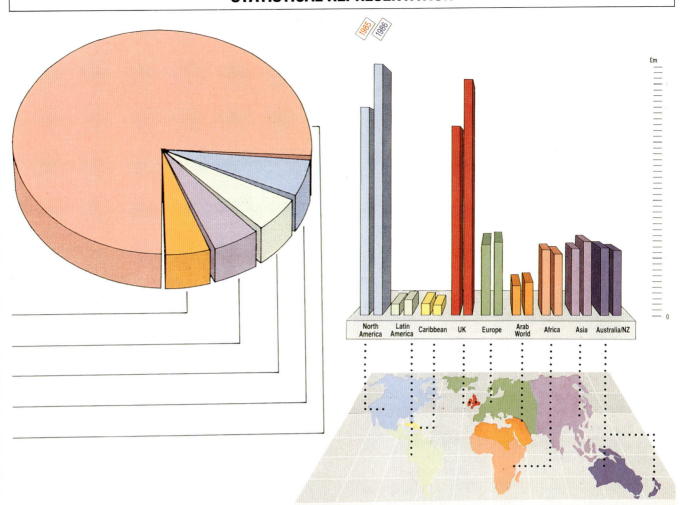

1985 1986

North America | Latin America | Caribbean | UK | Europe | Arab World | Africa | Asia | Australia/NZ

£m

SYMBOLS AND LOGOS

CERTIFICATION TRADE MARK

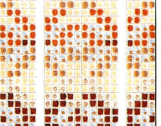

Size 40   Size 40
Size 34   Size 34
Size 37   Size 37

Add Bass
b)

Add Alto
c)

Add Bass
d)

INSTRUCTIONAL DRAWINGS

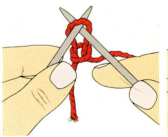

3 With the slip loop on your left-hand needle, insert your right-hand needle through the loop from front to back.

5 Draw up thé yarn through the slip loop to make a stitch.

4 Bring the yarn under and over your right-hand needle.

6 Place the stitch on your left-hand needle. Continue to make stitches drawing the yarn through the last stitch on your left-hand needle.

DEVELOPING YOUR OWN STYLE

The way you present your ideas affects the way others understand them.

Here are a few examples of how some pupils have used graphics to put over their ideas. There are many techniques of drawing and you should experiment to find out which suits you best. The most important thing is that your ideas should be clear and easy to understand.

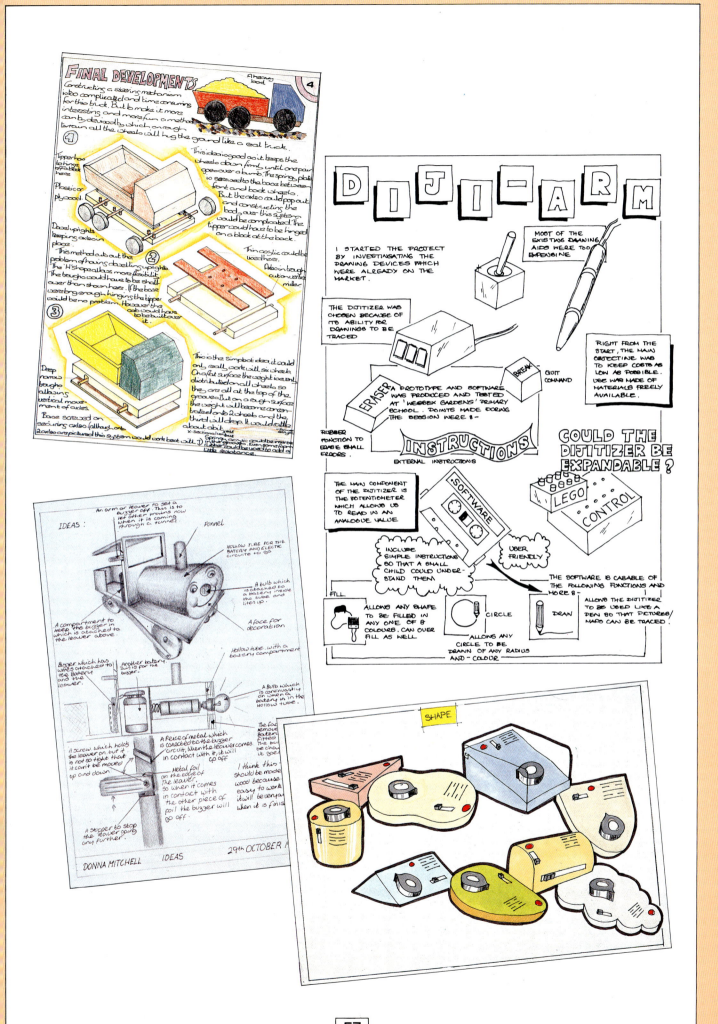

A PUPIL'S PROJECT

Susan Miles noticed that many parents used the handles of their baby buggies to carry shopping. This made the buggy unstable, so she tried to design a better way of carrying heavy loads.

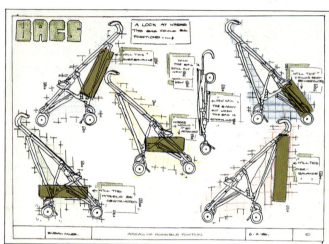

Susan's project led her to meetings with many experts in design and manufacturing. This gave her a good idea of the long road there was from idea to production and gave her the confidence to deal with real world problems.

HOW THE PROFESSIONALS DO IT

STRIDA

Whilst using public transport, Mark Saunders realised the limitations of existing folding bikes. This inspired him to begin designing the Strida as a low-cost, lightweight and easily collapsible alternative. Although retaining some of the basic principles, it is a fundamental reworking of traditional bicycle design.

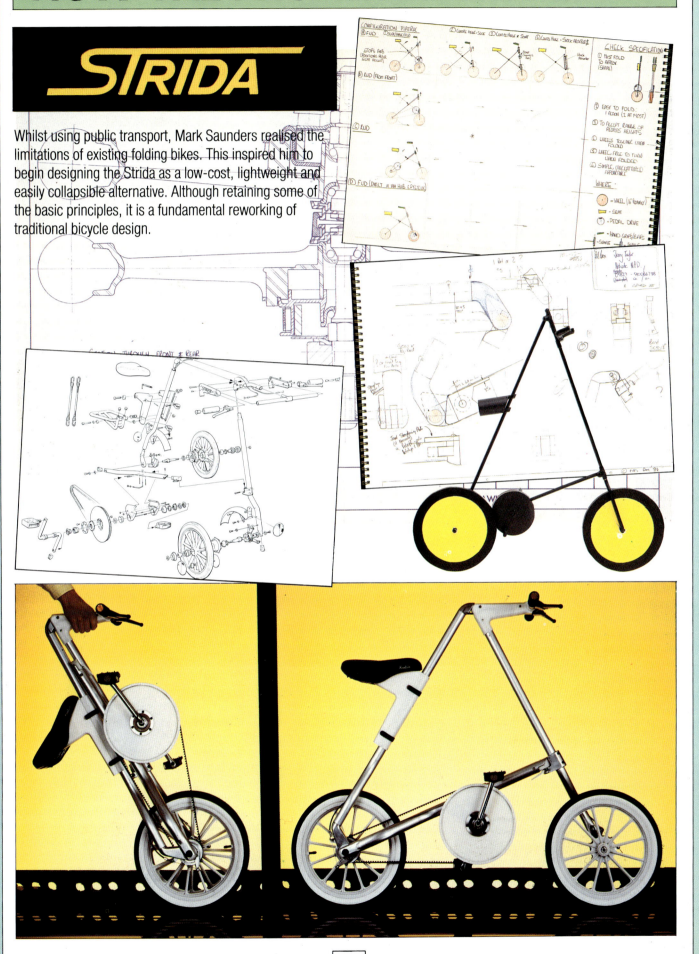

MODELLING AN IDEA

In CDT, modelling is always used to investigate form and function. The methods used can vary considerably depending on what is required.

NON-WORKING PROTOTYPES

These models are meant to look like the final product, but they do not need to function. They test aspects such as form, finish, colour selection, ergonomics, size etc. They may be full size or to scale.

MODELS FOR QUICK INVESTIGATION

These are not 'finished' ideas but help the designer to evaluate in three dimensions.

Pen

Strida

Calculators

Theatre set

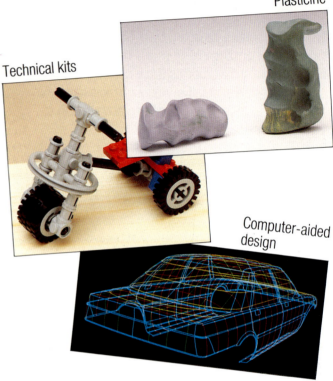

Plasticine

Technical kits

Computer-aided design

WORKING PROTOTYPES

These are used to test the final form and function of a design. In industry this stage would be followed by full-scale manufacture.

Torches

Timer

LEVEL C

INTRODUCTION TO LEVEL C

Level C projects demand greater skills than those in Levels A and B, and are the most open-ended. A growing awareness of how design and technology affects society will play a key role in your decision-making.

Broad situations are presented with guidelines of areas to research and analyse. Your research will draw on resources both inside and outside school; it is up to you to develop ideas in the directions suggested by your investigations.

As your ideas become more complex your design sheets will need to communicate them more effectively. Testing and modelling will become increasingly important in the development and evaluation of your designs.

What categories of magazines are available?
What kind of offers might they have?
What has been given as a 'freebie' in the past?
What size and quality were these free gifts?
How were they made?
Were they useful and did they have a general appeal?
How did they catch the customer's eye?

IDENTIFYING A SITUATION

WHAT PROBLEMS MIGHT THE FREEBIES SOLVE?

SORTING AND MEASURING

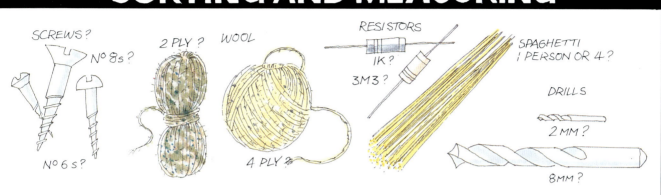

SCREWS?

Nº 8s?

Nº 6s?

2 PLY?

WOOL

4 PLY?

RESISTORS

1K?

3M3?

SPAGHETTI
1 PERSON OR 4?

DRILLS

2 MM?

8 MM?

HOBBIES

SNOOKER?

STAMPS?

2d

TROPICAL FISH?

GARDENING?

SPORTS?

COSMETICS

EYE SHADOW?

NAIL
VARNISH?

PONG

PERFUME
SAMPLES?

LIPSTICK?

MOTORCAR

DIP STICK
CLEANER?

ICE
SCRAPER?

DRIVE

KEY FOB?

WHAT OTHER POSSIBILITIES ARE THERE?

KITCHEN   LEISURE   SCHOOL   OFFICE   MUSIC   WHAT ELSE?

SPECIFICATION

After you have identified a situation and a possible idea for a 'freebie', write up a **specification** to guide you in your design work. Consider . . . . .

PIGEON FANCIER — FREE ½ OZ OF FEED — APPEAL

POTATO GROWER — FREE BALANCE WEIGHT — WEIGHT — 10 KG

FRANK NEWS FOR PHILATELISTS — Free — PENNY BLACK — ECONOMY

DIY WEEKLY — FREE PLANK OF WOOD — GRAPHICS — SIZE

DESIGNING

When you start your design ideas, bear in mind all those points in your specification, as well as ease and techniques of manufacture, packaging etc.
Evaluate each of your ideas and ask other people to evaluate them in order to decide which is the most suitable.

MODELLING

★ What materials would suit the design?
★ What tools and techniques would you use to model?
★ How can you achieve a 'professional' finish?

GRAPHICS AND PACKAGING

HOW CAN YOU ATTRACT THE EYE?

FREE

SPECIAL OFFER

FREE LIGHT ON DARK?

'STARBURSTS'? SPECIAL LETTERING STYLES?

HOW CAN YOU FIX ITEMS TO MAGAZINE COVERS?

DUCK BREEDER — BREED YOUR OWN MALLARD

WILL THE FREEBIE NEED PROTECTION FROM DAMAGE?

GARDEN — 3 PACKETS OF RARE ANNUALS

DOES THE FREEBIE NEED 'CONTAINING'?

GOOD HEALTH

Biscuits and sweets are one area of high-fat food often forgotten in dietary terms, yet sometimes eaten in great quantity.

High-fat foods contribute to heart disease and the general public are becoming increasingly aware of what they eat.

Even large food companies are now responding to this demand by marketing ranges of wholefood products. The way in which these are presented is an important part of their success.

RESEARCH

WHAT DO YOU EAT?

HOW HEALTHY IS IT?

HOW CAN YOU MAKE AN IMPROVEMENT IN YOUR DIET?

INFORMATION

COLLECTING AND RECORDING

DISPLAYING —

HOW CAN YOU SHOW THE RESULTS OF YOUR RESEARCH CLEARLY?

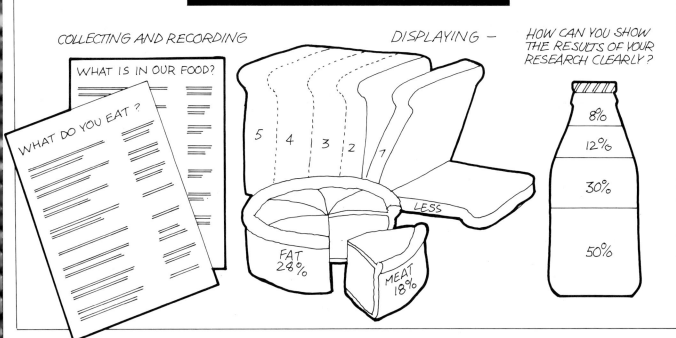

WHAT IS IN OUR FOOD?

WHAT DO YOU EAT?

5 4 3 2 1

LESS

FAT 28%

MEAT 18%

8%

12%

30%

50%

PRODUCT IDENTITY

Investigate how food manufacturers give their product a
'personality' by using shape, colour, texture, etc.

SWISSROLL

WEETABIX

DOUGHNUT

BOURBON

LIQUORICE ALLSORTS

HOOLA HOOPS

BATTENBERG

DESIGN · IDEAS

HOW CAN YOU GIVE YOUR PRODUCT AN IDENTITY?

SHAPE AND FORM

WHAT RECIPE WILL YOU NEED TO MAKE THESE IDEAS?

TEXTURE

HOW COULD YOU PROVIDE AN INTERESTING SURFACE ON YOUR DESIGN?

PRESSING
GRAINS?
NUTS?
CORN FLAKES?

COLOUR

WHAT NATURAL INGREDIENTS COULD PROVIDE COLOUR?

COCOA?

COCONUT?

DRIED FRUITS?

PATTERN

COULD THE FEATURE OF YOUR PRODUCT BE ITS PATTERN?

LATTICE?

PIPED?

APPLIED?

CUTTERS
– MAKING YOUR PRODUCT A UNIFORM SHAPE

SHEET METALS OR PLASTICS?

MILD STEEL OR BRONZE ROD?

PVC TUBING?

SHEET TEMPLATE?

WHICH MATERIALS ARE THE MOST SUITABLE FOR THE CUTTER?

MATERIAL	YES	NO	REASON
TINPLATE			
ACRYLIC			
SOFTWOOD			
PVC			

PACKAGING

WHAT IS ITS PURPOSE?

HOW CAN THE PRODUCT BE PACKED?

WHAT MATERIALS AND TECHNIQUES CAN YOU USE?

HOW CAN YOU AVOID DAMAGE?

WHERE CAN YOU USE GRAPHICS?

TESTING AND EVALUATION

CONSUMER RESEARCH

How, by comparative testing can you find out consumer reaction?

What will you test? – Appearance? Taste? Contents? Packaging?

Should you consider reactions of different groups of people – age, gender, ethnic background?

How can you display your results?

DASHBOARD DESIGN

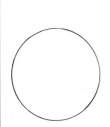
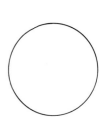

The speed and clarity of the information given to the driver of any vehicle is very important. Sometimes it can save life, or be the cause of a serious accident.

The way we see things is affected by many factors and good design considers all situations in which the display might be used.

Investigate existing systems and propose a design for the dashboard of a specific type of vehicle.

EXISTING SYSTEMS

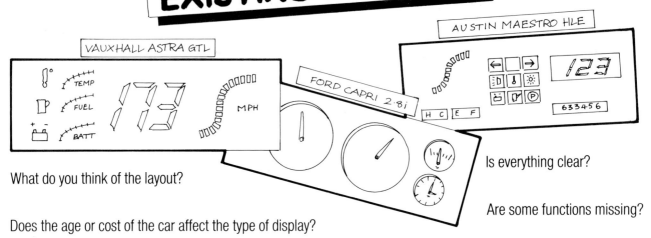

VAUXHALL ASTRA GTL

AUSTIN MAESTRO HLE

FORD CAPRI 2.8i

What do you think of the layout?

Does the age or cost of the car affect the type of display?

Is everything clear?

Are some functions missing?

'FUNCTIONS'

WHAT INFORMATION IS ESSENTIAL AND WHAT IS OPTIONAL?

Lights?

Indicators?

Handbrake?

Revs per minute?

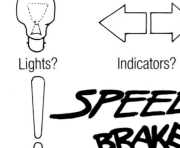

Hazard Warning?

SPEED
BRAKES

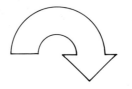

Tape/Radio/CD?

IGNITION ?
FUEL ?   OIL ?

Volume?

Temperature?

SYSTEMS MODING

What information is needed at particular times?
★ When starting?
★ When travelling at 70 MPH?
★ At night?
★ In cold weather?
   etc.

What information should carry the highest priority at each of these times?

COMMUNICATING INFORMATION

HOW IS INFORMATION PUT OVER WITH INDICATORS AND CONTROLS?

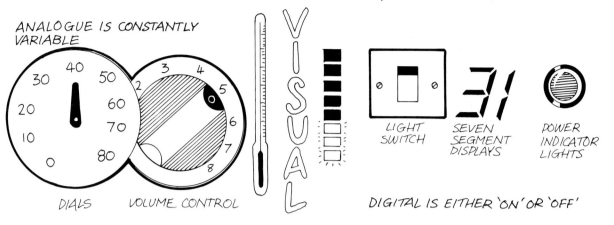

ANALOGUE IS CONSTANTLY VARIABLE

V I S U A L

DIALS

VOLUME CONTROL

LIGHT SWITCH

SEVEN SEGMENT DISPLAYS

POWER INDICATOR LIGHTS

DIGITAL IS EITHER 'ON' OR 'OFF'

AUDIBLE

WHERE IS SOUND USED TO ATTRACT ATTENTION?

FIRE BELLS

CAR HORN

WHISTLING KETTLE

POLICE SIREN

IDEOGRAMS
HOW CAN YOU DESCRIBE A FUNCTION USING A SYMBOL?

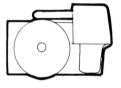

IGNITION?

HAZARD WARNING?

LIGHTS?

WATER TEMPERATURE?

OIL LEVEL?

TESTING AND EVALUATION

MAKE A 'MOCK-UP' FROM THE RESULTS OF YOUR INVESTIGATION...

NAME	DIG.	ANA.
ANITA	0.5	0.6
BIMAL	0.4	0.6
AZRA	0.5	0.8
JOHN	0.6	0.7

HOW CAN YOU TEST AND RECORD WHICH DISPLAY CAN BE IDENTIFIED THE QUICKEST?

WHAT CONCLUSIONS CAN YOU MAKE FROM YOUR TESTS?

HOW SUCCESSFUL IS THE LAYOUT? HOW WOULD EACH DISPLAY FUNCTION?

PRODUCT DESIGN

Despite the fact that many products are functional in their nature, our attitude towards their appearance strongly influences our decision when we come to purchase them.

Designers are conscious of this tendency and use it to their advantage when styling a product.

Investigate how products are aimed at 'target groups' (e.g. teenagers, sportspeople, the fashion-conscious, executives etc.) and use your results to help your own product design.

WHAT PRODUCTS CAN YOU INVESTIGATE?

WATCHES

PENS

BAGS AND CASES

TOILETRIES

ANYTHING ELSE?

RESEARCH AND ANALYSIS

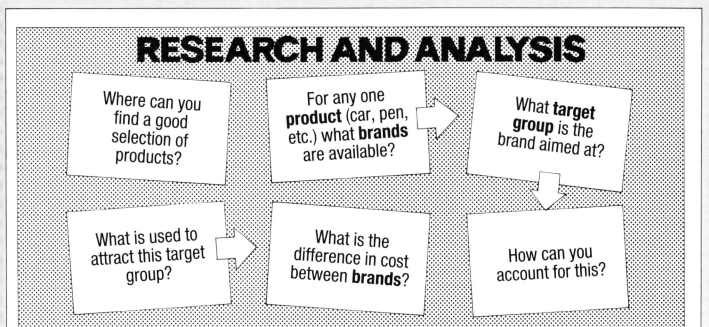

Where can you find a good selection of products?

For any one **product** (car, pen, etc.) what **brands** are available?

What **target group** is the brand aimed at?

What is used to attract this target group?

What is the difference in cost between **brands**?

How can you account for this?

MANUFACTURING

Is the product made up from several parts?
What function does each perform?
How are the parts made?
How does the manufacturing process affect the product's price?
Can the product be made in different ways?

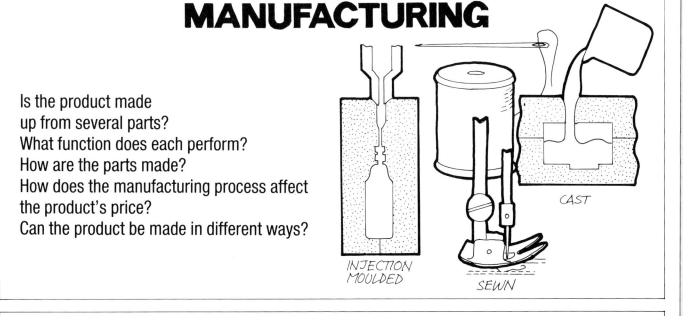

INJECTION MOULDED

SEWN

CAST

RECORDING AND EVALUATION

★ How can you record your research?
★ What factors do you think influence people when they are buying a product?
★ What have you discovered that could help your own design work?
★ Which brands do you think are the most successful and why?

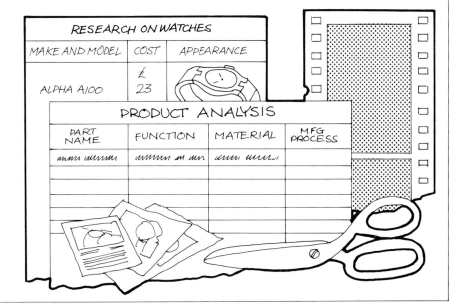

RESEARCH ON WATCHES

MAKE AND MODEL	COST	APPEARANCE
ALPHA A100	£ 23	

PRODUCT ANALYSIS

PART NAME	FUNCTION	MATERIAL	MFG PROCESS

CALCULATOR STYLING

For most people the more complex functions of 'mathematical' or 'scientific' calculators are unnecessary. Investigate and record existing products and consider this information in your product design.

TYPICAL LAYOUT OF EXISTING CALCULATOR

ANALYSIS AND RESEARCH

★ When and why do people use calculators?
★ What 'functions' are used?
★ What is on the market already?
★ What colours and finishes are used on existing designs?
★ Who do these designs appeal to?
★ How important are size, clarity, ease of use and appearance?
★ What kind of packaging is used?
★ Does the layout of the keys need to be like the one shown?
★ What kind of displays are available?
★ How are calculators powered?

KEYS

WHAT SHAPE AND SIZE SHOULD THEY BE ?

WILL THEY STAND OUT FROM THE SURFACE?

- OR COULD THEY BE TOUCH PADS?

HOW FAR APART ?

- OR MEMBRANE SWITCHES ?

WHAT ABOUT KINAESTHETIC FEEDBACK !!?

POWER SOURCE

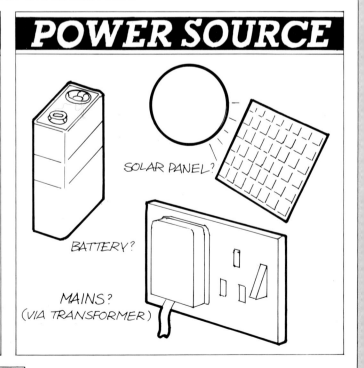

SOLAR PANEL?

BATTERY?

MAINS? (VIA TRANSFORMER)

CAN THE PRODUCT BE GIVEN
AN INTERESTING APPEARANCE?

DESIGN IDEAS

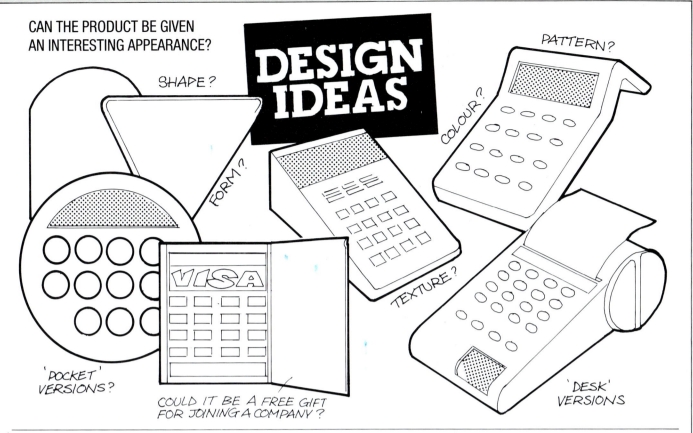

SHAPE?

PATTERN?

COLOUR?

FORM?

TEXTURE?

'POCKET' VERSIONS?

COULD IT BE A FREE GIFT
FOR JOINING A COMPANY?

'DESK' VERSIONS

MODELLING

WHAT MATERIALS AND PROCESSES ARE AVAILABLE?
HOW CAN YOU ACHIEVE A 'PROFESSIONAL' FINISH?

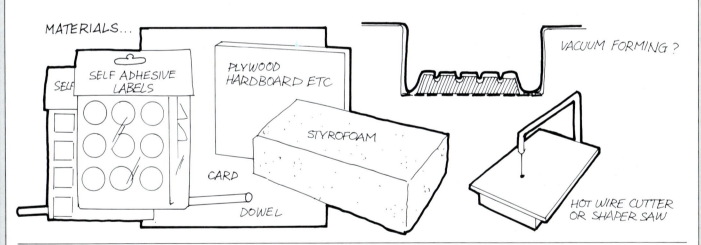

MATERIALS...

SELF ADHESIVE LABELS

PLYWOOD HARDBOARD ETC

VACUUM FORMING?

STYROFOAM

CARD

DOWEL

HOT WIRE CUTTER
OR SHAPER SAW

DISPLAY AND PACKAGING

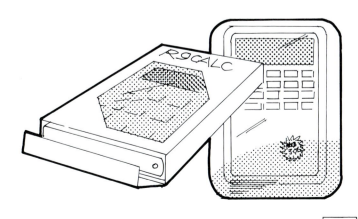

★ What is the purpose of packaging?
★ How can packaging be used as display?
★ How can you attract the customer's attention through packaging?
★ How can the design be displayed on a sales counter?
★ Is the problem similar to book display?

Designer Pens

Watches have become popular fashion items, opening a new market for manufacturers. Could the sale of pens be increased in the same way by better design?

Research and Analysis

- ★ What is on the market already and what does it cost?
- ★ What 'target group' does the product try to attract and who will you aim for?
- ★ What type of packaging is used and how effective is it?
- ★ How are you limited in what you can do?

Materials and Processes

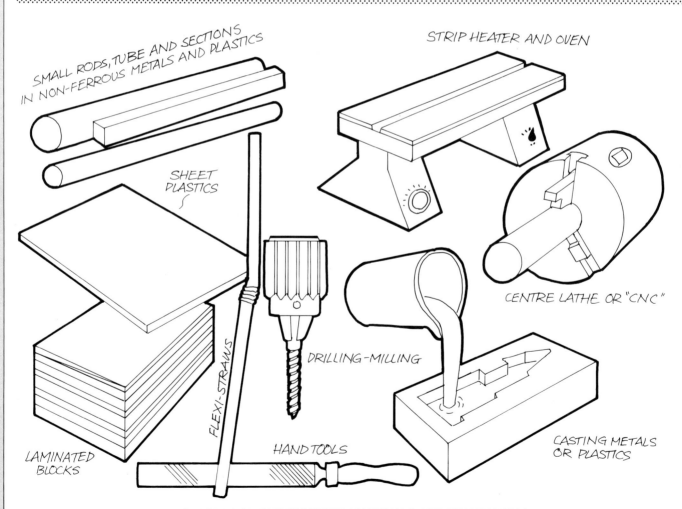

SMALL RODS, TUBE AND SECTIONS IN NON-FERROUS METALS AND PLASTICS

STRIP HEATER AND OVEN

SHEET PLASTICS

CENTRE LATHE OR "CNC"

FLEXI-STRAWS

DRILLING-MILLING

CASTING METALS OR PLASTICS

LAMINATED BLOCKS

HAND TOOLS

HOW CAN YOU PUT TOGETHER MATERIALS AND TECHNIQUES?

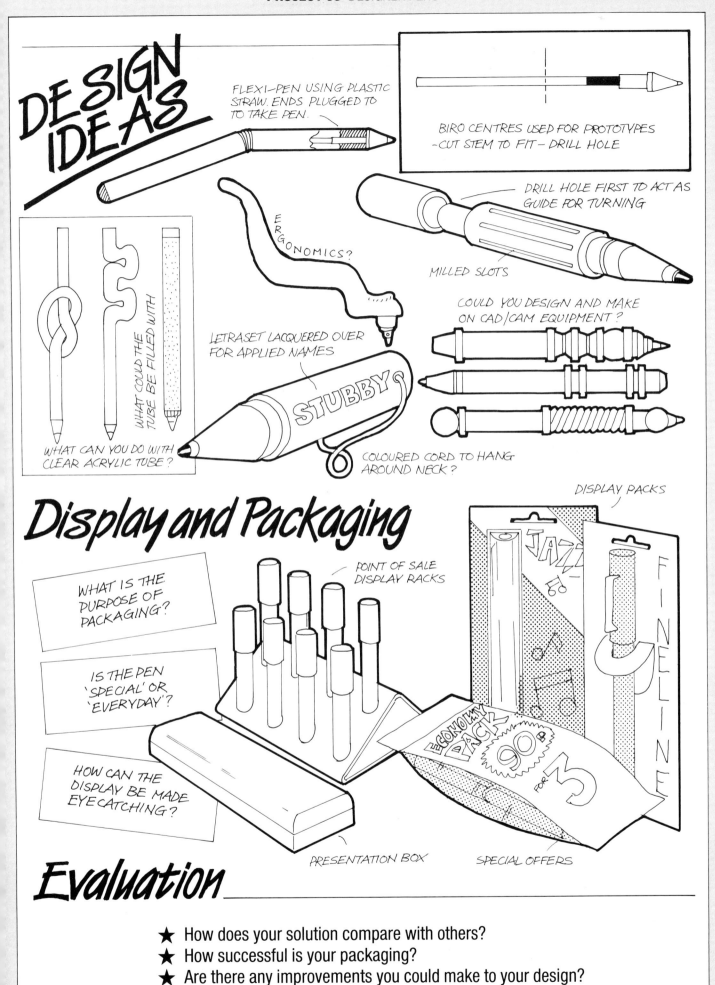

DESIGN IDEAS

FLEXI-PEN USING PLASTIC STRAW. ENDS PLUGGED TO TO TAKE PEN.

BIRO CENTRES USED FOR PROTOTYPES -CUT STEM TO FIT - DRILL HOLE

DRILL HOLE FIRST TO ACT AS GUIDE FOR TURNING

MILLED SLOTS

ERGONOMICS?

COULD YOU DESIGN AND MAKE ON CAD/CAM EQUIPMENT?

LETRASET LACQUERED OVER FOR APPLIED NAMES

STUBBY

WHAT COULD THE TUBE BE FILLED WITH

WHAT CAN YOU DO WITH CLEAR ACRYLIC TUBE?

COLOURED CORD TO HANG AROUND NECK?

Display and Packaging

WHAT IS THE PURPOSE OF PACKAGING?

IS THE PEN 'SPECIAL' OR 'EVERYDAY'?

HOW CAN THE DISPLAY BE MADE EYE CATCHING?

POINT OF SALE DISPLAY RACKS

DISPLAY PACKS

JAZZ

FINELINE

ECONOMY PACK 90p 3 FOR

PRESENTATION BOX

SPECIAL OFFERS

Evaluation

★ How does your solution compare with others?
★ How successful is your packaging?
★ Are there any improvements you could make to your design?

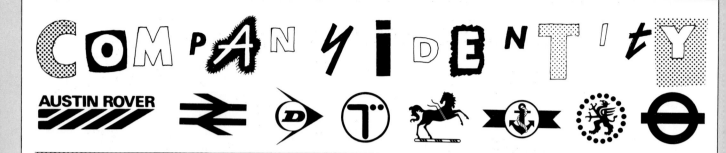

COMPANY IDENTITY

All companies are in the business of selling. Some sell objects, some sell services, and to do this they are also selling themselves. Each company develops an 'image' or an identity which separates it from its competitors. This makes it instantly recognisable to the public and is used to promote the company and its products.

In order to increase their share of the market, a company needs both good products and a strong and attractive image to keep it in the public eye.

HOW DO COMPANIES DEVELOP AN IDENTITY?

NAME
LOGO
▷ ▷ ▷ ▷
RESEARCH
ADVERTISING
MAGAZINE

WHAT IMAGE DO THEY TRY TO PRESENT?

RECORDING INFORMATION

COMPANY NAME	LETTERING / LOGO	ANALYSIS / COMMENT
THE WHAKIE HAIR COMPANY	 THE WHAKIE HAIR COMPANY	A NEW LOOK HAIR STUDIO– STYLE– YOUTHFUL– INDIVIDUAL– WAY OUT– APPEALING TO 16–18 YEARS
AIR CANADA	AIR CANADA A BREATH OF FRESH AIR	MAPLE LEAVES–SYMBOL OF CANADA FLOATING–GRACEFUL– DISAPPEARING INTO THE SKY LIKE A PLANE EASY, RESTFUL FLIGHT

HOW DO THEY CREATE AN INTEREST?

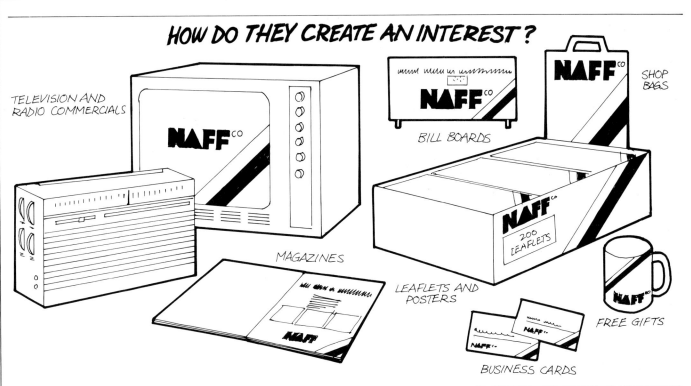

TELEVISION AND RADIO COMMERCIALS

BILL BOARDS

SHOP BAGS

MAGAZINES

LEAFLETS AND POSTERS

FREE GIFTS

BUSINESS CARDS

ANALYSIS

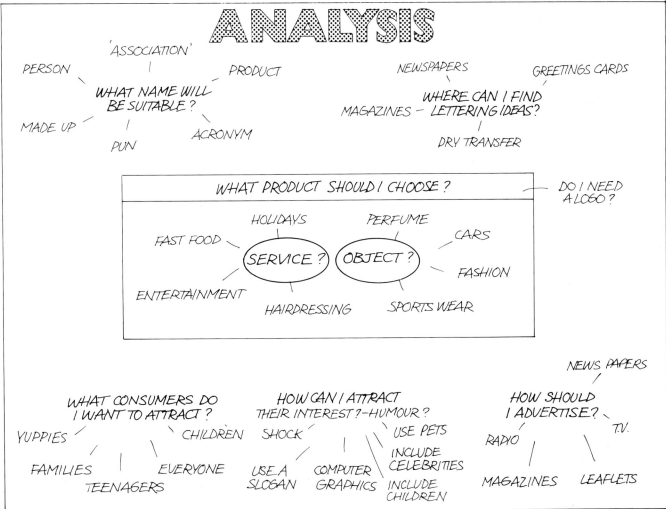

PERSON

'ASSOCIATION'

PRODUCT

WHAT NAME WILL BE SUITABLE?

MADE UP

PUN

ACRONYM

NEWSPAPERS

GREETINGS CARDS

WHERE CAN I FIND LETTERING IDEAS?

MAGAZINES

DRY TRANSFER

WHAT PRODUCT SHOULD I CHOOSE?

DO I NEED A LOGO?

HOLIDAYS

PERFUME

FAST FOOD

CARS

SERVICE?

OBJECT?

FASHION

ENTERTAINMENT

HAIRDRESSING

SPORTS WEAR

WHAT CONSUMERS DO I WANT TO ATTRACT?

YUPPIES

CHILDREN

FAMILIES

EVERYONE

TEENAGERS

HOW CAN I ATTRACT THEIR INTEREST? - HUMOUR?

SHOCK

USE PETS

USE A SLOGAN

COMPUTER GRAPHICS

INCLUDE CELEBRITIES

INCLUDE CHILDREN

HOW SHOULD I ADVERTISE?

NEWS PAPERS

RADIO

T.V.

MAGAZINES

LEAFLETS

SPECIFICATION

What product is your company going to sell?
What image do you wish to present?
How are you going to promote yourself?

DESIGN IDEAS

IDEAS SHEETS

KEEPING IDEAS TOGETHER

JOLLY JOE'S
Joe's
Joe's CAFE
J's Restaurant
NAMES

JOE'S CAFE

LOGOS

CHOSEN IDEA

FOLDERS ?

PLASTIC
CARD
PAPER
ENVELOPE ?
WALLET ?
BINDER ?

WHICH IDEAS BEST SUIT YOUR 'IMAGE'?

Private barbecues catered for
LAMS
WEST MALAYSIAN FOOD CO.

GRUNTS
12 MAIDEN LANE
LONDON WC2
01 379 0DZ

RISTORANTE · PENSIONE
"LE ONDE"
di ALLA ROBERTO

BUSINESS CARDS

What information is essential?

How can you make your card worth keeping?

ADVERTISING LEAFLET

What information is essential?
Why are maps often included?
– How can you make the location clear?

How will people travel to your premises?
– public transport?
– private transport?

MONTAGU ROAD
EDMONTON
ROTTERIA
NORTH CIRCULAR ROAD
CAR PARK
CHINGFORD
PARK HERE
A MILE ACROSS
STACEY AV'S
STACEY AV'S
PRINCES ROAD
CAR PARK
YOU ARE HERE
MAP SHOWS TWO PLACES TO PARK

78

NEWSPAPER ADVERTISMENTS

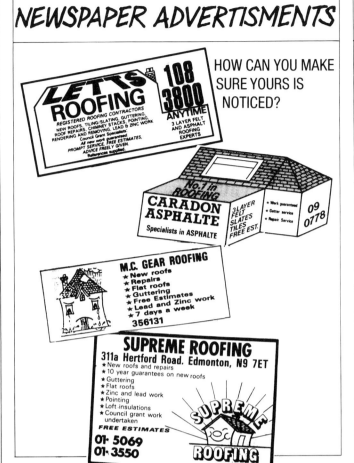

HOW CAN YOU MAKE SURE YOURS IS NOTICED?

ADVERTISING GIVEAWAYS

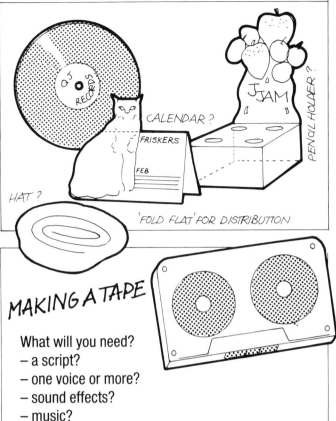

CALENDAR?

PENCIL HOLDER?

HAT?

'FOLD FLAT' FOR DISTRIBUTION

MAKING A TAPE

What will you need?
– a script?
– one voice or more?
– sound effects?
– music?
How will you get both product and image over to the public?

T.V. ADVERTISING :

PRODUCING A STORYBOARD

FRISKERS: Image – Friskers is a food for home loving cats and caring owners.

'TOOTS! TOOTS'

TOOTS ALWAYS PREFERS FRISKERS!

Can you follow the 'story'?   Do you need extra 'frames'?   Will you need to write the story first?

Evaluation MARKET RESEARCH

Have you made people interested?
Will they remember your company and its products?

Schools are often involved in fund-raising.
Schools themselves sometimes need to raise money for large projects, and pupils can participate in promotion and collection.
    Many pupils are also concerned about issues and organisations outside school life, and they can show support by taking or sharing responsibility for a campaign in their own school.

RESEARCH

HOW DO ORGANISATIONS RAISE FUNDS?

HELP!

PUBLICITY

MONEY COLLECTION

CHARITY SHOPS

EVENTS
JUMBLE SALE
CONCERTS
SPONSORSHIPS

1988

CATALOGUE SHOPPING

PLEASE GIVE GENEROUSLY

ENVELOPES FOR DOOR TO DOOR COLLECTIONS

STICKERS, FLAGS ETC.

WHAT FUND-RAISING HAS YOUR SCHOOL BEEN INVOLVED IN DURING THE LAST YEAR?

DISASTERS, EARTHQUAKES, DROUGHT

FAMINE

SCHOOL PROJECT

CHARITIES

★ How successful were the campaigns?

★ Can campaigning be made more effective?

ANALYSIS

FUND RAISING

What publicity is needed for a fund-raising campaign in school?

How much money can be spent?

Can everything be produced at school?

What materials will be needed?

SPECIFICATION

What containers will be needed?

What or whom should the campaign raise money for? Should a 'target' be set? What items need to be produced for publicity and collection?

DESIGN IDEAS

HOW CAN YOU DEVELOP AN IDENTITY TO UNIFY YOUR CAMPAIGN?

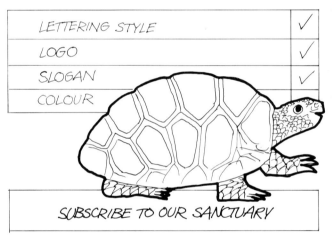

LETTERING STYLE	✓
LOGO	✓
SLOGAN	✓
COLOUR	

SUBSCRIBE TO OUR SANCTUARY

HOW CAN YOU PUBLICISE YOUR CAMPAIGN?

POSTERS

?

WATCH THIS SPACE!

MYSTERY?

LEAFLETS

DO YOU KNOW?
200 TORTOISES A MONTH ARE ABANDONED? REPORTS SUGGEST UNWANTED TORTOISES ARE SOLD FOR SOUP AND THEIR SHELLS ARE USED FOR ASHTRAYS?

QUESTIONS?

WANTED

SANCTUARY FOR THIS TORTOISE

SHOCK?

WILL A 'TOTALS DISPLAY' HELP TO RAISE MORE MONEY?

COMPETITION

X	£1 £1 £1 £1
Y	£1 £1 £1 £1 £1
Z	£1 £1 £1 £1 £1 £1 £1

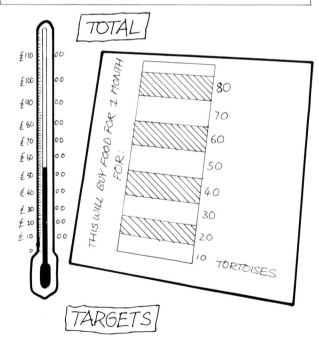

TOTAL

£110 00
£100 00
£90 00
£80 00
£70 00
£60 00
£50 00
£40 00
£30 00
£20 00
£10 00
0

THIS WILL BUY FOOD FOR 1 MONTH FOR:

80
70
60
50
40
30
20
10 TORTOISES

TARGETS

Is the information easy to understand?

What scale should be used?

WHAT ITEMS WILL BE NEEDED?

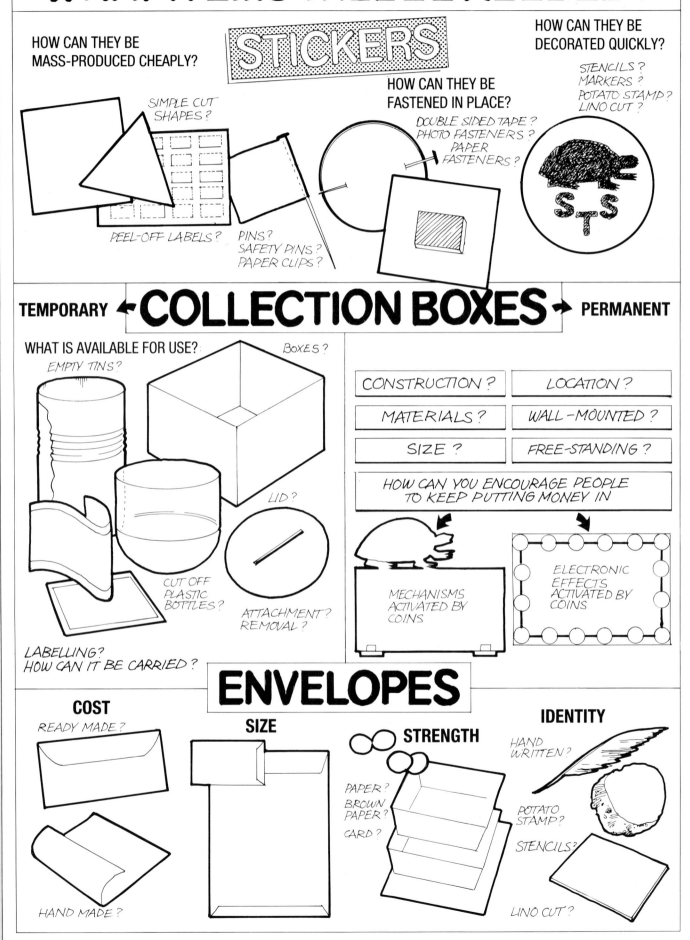

HOW CAN THEY BE
MASS-PRODUCED CHEAPLY?

STICKERS

HOW CAN THEY BE
DECORATED QUICKLY?

STENCILS ?
MARKERS ?
POTATO STAMP ?
LINO CUT ?

SIMPLE CUT
SHAPES ?

HOW CAN THEY BE
FASTENED IN PLACE?

DOUBLE SIDED TAPE ?
PHOTO FASTENERS ?
PAPER FASTENERS ?

S.T.S

PEEL-OFF LABELS ?

PINS ?
SAFETY PINS ?
PAPER CLIPS ?

TEMPORARY ← COLLECTION BOXES → PERMANENT

WHAT IS AVAILABLE FOR USE?

BOXES ?

EMPTY TINS ?

CONSTRUCTION ?	LOCATION ?
MATERIALS ?	WALL-MOUNTED ?
SIZE ?	FREE-STANDING ?

HOW CAN YOU ENCOURAGE PEOPLE
TO KEEP PUTTING MONEY IN

LID ?

CUT OFF
PLASTIC
BOTTLES ?

ATTACHMENT ?
REMOVAL ?

MECHANISMS
ACTIVATED BY
COINS

ELECTRONIC
EFFECTS
ACTIVATED BY
COINS

LABELLING ?
HOW CAN IT BE CARRIED ?

ENVELOPES

COST

READY MADE ?

SIZE

STRENGTH

IDENTITY

HAND
WRITTEN ?

PAPER ?
BROWN
PAPER ?
CARD ?

POTATO
STAMP ?

STENCILS ?

HAND MADE ?

LINO CUT ?

Being aware of threats to our safety is an essential part of modern life, yet we quickly become used to dangers and begin to be careless. However, a forgetful moment can result in suffering, or even death. How can we use our CDT skills to improve people's awareness of safety?

WHAT SITUATIONS ARE POTENTIALLY DANGEROUS?

RESEARCH

WHAT ITEMS ARE POTENTIALLY DANGEROUS?

WORK?

STREET?

HOME?

COUNTRYSIDE?

SCHOOL

MEDICINES

DOSAGE

CHEMICALS

KILL INSECTICIDE

TOYS

ELECTRICAL GOODS

TOOLS

LITTER

WARNINGS? HOW ARE WE WARNED?

SIGNS

POSTERS

AIDS

WATCH OUT!

SYMBOLS

WHICH REPORT

HEALTH AND SAFETY

SURVEYS

COLLECTING AND RECORDING INFORMATION

INTERVIEWS

JAMES AGE 10 ACCIDENTS

1 ....

CAUSE OF ACCIDENTS AT HOME

1 ......
2 ......

SCHOOL PHOTOGRAPHS:

BIOLOGY LAB

CDT ROOM

PLAN: MY KITCHEN

BRIEF

Design and make a 2D or 3D display that you think will improve people's awareness of the dangers of a particular situation or item.

▼

ANALYSIS

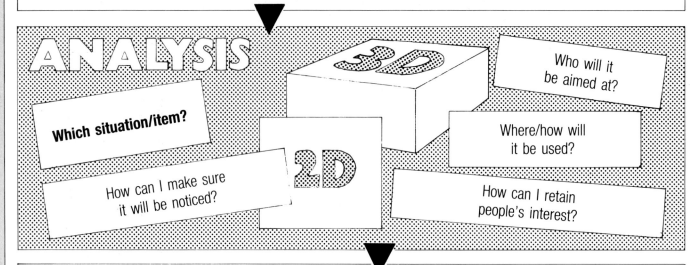

Who will it be aimed at?

Which situation/item?

How can I make sure it will be noticed?

Where/how will it be used?

How can I retain people's interest?

▼

SPECIFICATION

"STATE EXACTLY WHAT YOU INTEND TO MAKE".

DEVELOPING IDEAS

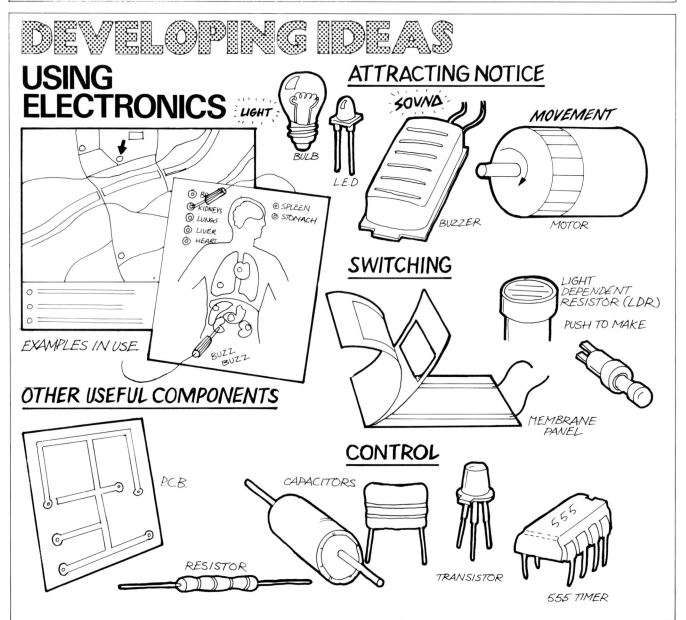

USING ELECTRONICS

LIGHT

BULB

L.E.D

EXAMPLES IN USE

BUZZ BUZZ

BR...
KIDNEYS
LUNGS
LIVER
HEART
SPLEEN
STOMACH

ATTRACTING NOTICE

SOUND

BUZZER

MOVEMENT

MOTOR

SWITCHING

LIGHT DEPENDENT RESISTOR (LDR)

PUSH TO MAKE

MEMBRANE PANEL

OTHER USEFUL COMPONENTS

P.C.B.

CONTROL

CAPACITORS

RESISTOR

TRANSISTOR

555 TIMER

USING ANIMATION

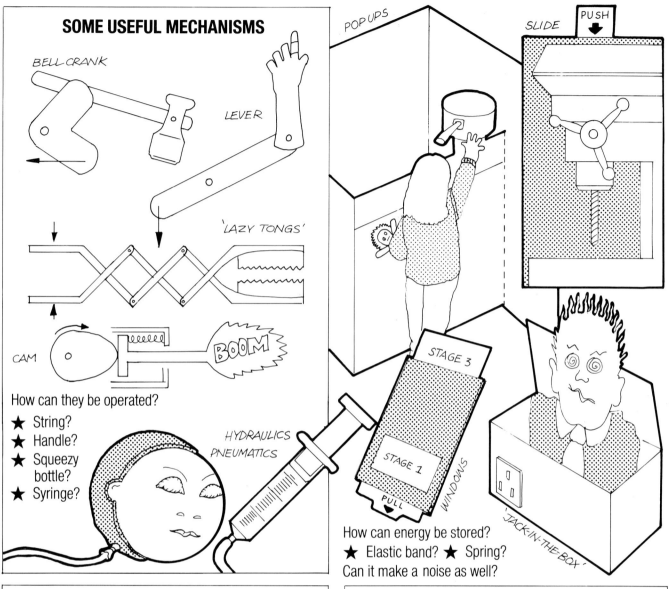

SOME USEFUL MECHANISMS

BELL CRANK

LEVER

'LAZY TONGS'

CAM

BOOM

How can they be operated?

★ String?
★ Handle?
★ Squeezy bottle?
★ Syringe?

HYDRAULICS
PNEUMATICS

POP UPS

SLIDE

PUSH

STAGE 3

STAGE 1

PULL

WINDOWS

'JACK-IN-THE-BOX'

How can energy be stored?
★ Elastic band? ★ Spring?
Can it make a noise as well?

WHAT MATERIALS WILL BE USED?

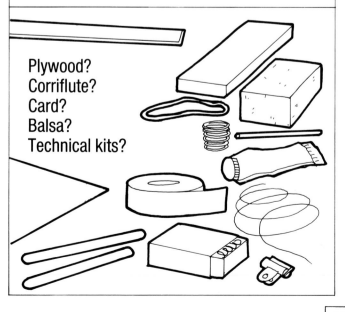

Plywood?
Corriflute?
Card?
Balsa?
Technical kits?

EVALUATION

BY QUESTIONNAIRE

NAME:
1 WHAT DO YOU THINK OF MY DISPLAY?

BY OBSERVATION?

NO. OF ACCIDENTS

Nº OF PEOPLE USING

BY COLLECTING DATA?

LUNAR ROVER

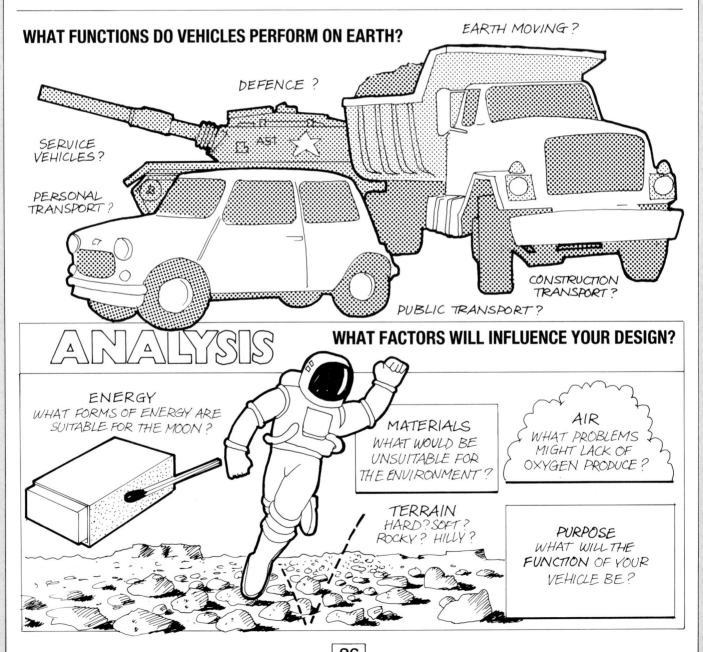

On earth, wheeled vehicles have evolved to suit our needs and climatic conditions. On the moon things are very different and the vehicles used in this environment will need to be carefully designed. Design and model a vehicle which could fulfil a particular need on the moon.

WHAT FUNCTIONS DO VEHICLES PERFORM ON EARTH?

EARTH MOVING ?

DEFENCE ?

SERVICE VEHICLES ?

PERSONAL TRANSPORT ?

CONSTRUCTION TRANSPORT ?

PUBLIC TRANSPORT ?

ANALYSIS

WHAT FACTORS WILL INFLUENCE YOUR DESIGN?

ENERGY
WHAT FORMS OF ENERGY ARE SUITABLE FOR THE MOON ?

MATERIALS
WHAT WOULD BE UNSUITABLE FOR THE ENVIRONMENT ?

AIR
WHAT PROBLEMS MIGHT LACK OF OXYGEN PRODUCE ?

TERRAIN
HARD? SOFT? ROCKY? HILLY ?

PURPOSE
WHAT WILL THE FUNCTION OF YOUR VEHICLE BE?

ENERGY

WHAT TYPES OF MOTOR COULD BE USED ON THE MOON?

WHAT FORMS OF ENERGY ARE AVAILABLE ON THE MOON?

WHAT ENERGY FORMS WOULD BE USED BY HUMANS ON THE MOON?

WATER?

WIND?

COAL?

OIL?

INTERNAL COMBUSTION?

RUBBER BAND? ELECTRIC?

CLOCKWORK?

STEAM?

MATERIALS

LEAD BALLOON

WHAT MATERIALS WOULD BE MOST SUITABLE FOR TAKING TO THE MOON?

WHAT CONDITIONS ON THE MOON WOULD AFFECT YOUR CHOICE OF MATERIAL?
HOW DOES THE TASK AFFECT YOUR CHOICE?

HOW USEFUL ARE MATERIALS COMMONLY USED ON EARTH?
• WOOD?   • CERAMICS?
• METALS?   • TEXTILES?
• PLASTICS? • GLASS?

CONTROL

HOW WOULD CONDITIONS ON THE MOON AFFECT THE WAY HUMANS COULD WORK? HOW COULD THIS BE OVERCOME?

RADIATION?

WEIGHT?

TEMPERATURE?

LIGHT?

HOW CAN MACHINES BE CONTROLLED AT A DISTANCE?

DESIGN DEVELOPMENT

BREAK YOUR PROBLEM DOWN INTO SMALLER 'SUB- PROBLEMS'

CHASSIS

WILL THIS BE A FLAT BASE OR FRAME CONSTRUCTION?

HOW CAN YOU STRENGTHEN A FRAMEWORK CHASSIS?

SHOULD IT BE ARTICULATED?

SUPERSTRUCTURE

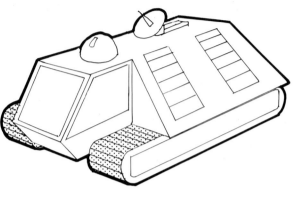

DOES THE VEHICLE NEED TO BE PROTECTED FROM THE ELEMENTS?

SPEED

WHICH IS MOST IMPORTANT, SPEED OR STRENGTH?

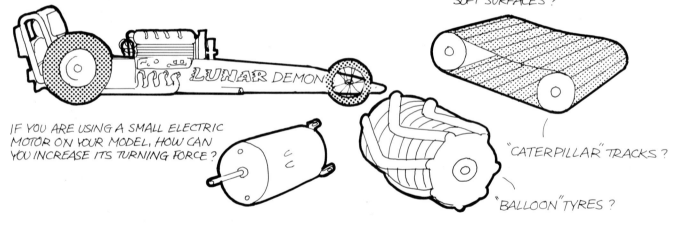

LUNAR DEMON

IF YOU ARE USING A SMALL ELECTRIC MOTOR ON YOUR MODEL, HOW CAN YOU INCREASE ITS TURNING FORCE?

TRACTION

HOW WOULD YOU VEHICLE TRAVEL OVER ROUGH OR SOFT SURFACES?

"CATERPILLAR" TRACKS?

"BALLOON" TYRES?

TRANSMISSION

GEARS

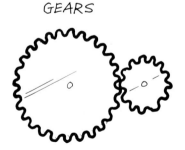

WHAT RATIO WOULD YOU WANT BETWEEN THE MOTOR AND WHEELS?

PULLEYS

WHAT SIZE PULLEYS WOULD YOU USE?

WHAT WOULD YOU USE FOR A BELT?

CHAIN AND SPROCKET

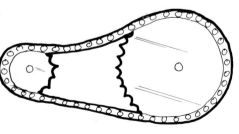

WHAT ADVANTAGE WOULD CHAIN AND SPROCKET HAVE OVER GEARS AND PULLEYS?

MODELLING

PUTTING YOUR IDEAS INTO '3D'

WILL IT BE A 'WORKING' OR A 'DISPLAY' MODEL?

WHAT MATERIALS CAN YOU USE?

TECHNICAL KITS?

STYROFOAM?

CARDBOARD DEVELOPMENTS?

'SALVAGED' WHEELS FROM TOYS?

'PLAWCO' OR WIRE FRAME?

'STICK AND GUSSET'?

'CORRIFLUTE' CUT TO MAKE CATERPILLAR TRACKS?

HOW CAN YOU CONTROL YOUR MODEL?

SMALL MOTORS?

'HYDRAULICS'?

COMPUTER CONTROL?

RECORDING YOUR RESULTS

Can you make your model look as if it were on the moon?

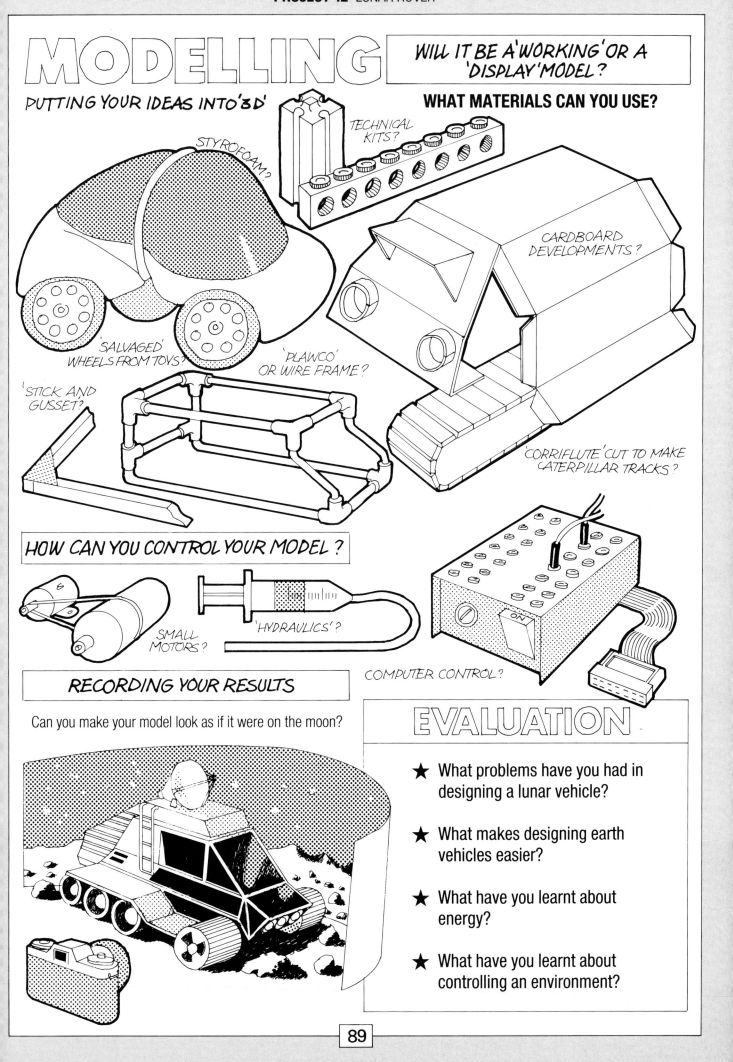

EVALUATION

★ What problems have you had in designing a lunar vehicle?

★ What makes designing earth vehicles easier?

★ What have you learnt about energy?

★ What have you learnt about controlling an environment?

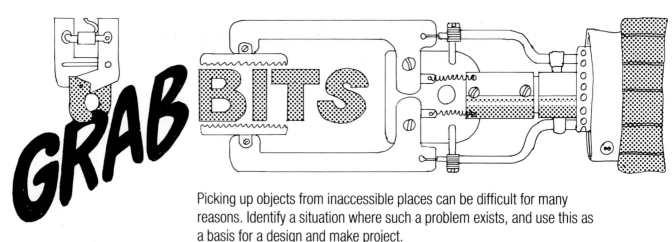

GRAB BITS

Picking up objects from inaccessible places can be difficult for many reasons. Identify a situation where such a problem exists, and use this as a basis for a design and make project.

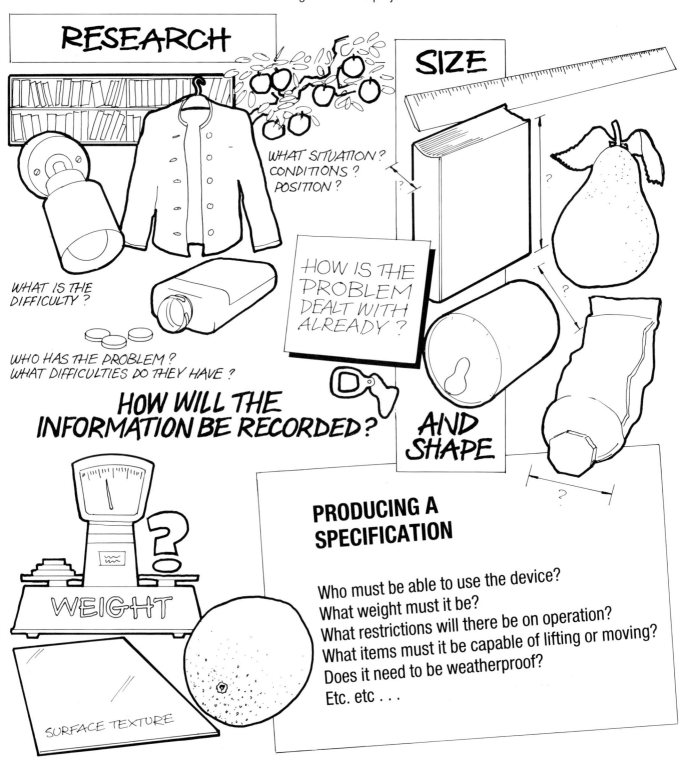

RESEARCH

WHAT SITUATION?
CONDITIONS?
POSITION?

WHAT IS THE DIFFICULTY?

WHO HAS THE PROBLEM?
WHAT DIFFICULTIES DO THEY HAVE?

HOW IS THE PROBLEM DEALT WITH ALREADY?

HOW WILL THE INFORMATION BE RECORDED?

WEIGHT

SURFACE TEXTURE

SIZE

AND SHAPE

PRODUCING A SPECIFICATION

Who must be able to use the device?
What weight must it be?
What restrictions will there be on operation?
What items must it be capable of lifting or moving?
Does it need to be weatherproof?
Etc. etc . . .

DEVELOPING IDEAS

MECHANISMS

SOME EXAMPLES OF LEVERS AND LINKAGES IN USE...

CLOTHES PEG: WHICH CLASS OF LEVER IS IN OPERATION HERE?

"LAZY TONGS" OR PENTOGRAPH LINKAGE

COULD LEVERS OR LINKAGES BE INCORPORATED?

CLASS
1
2
3

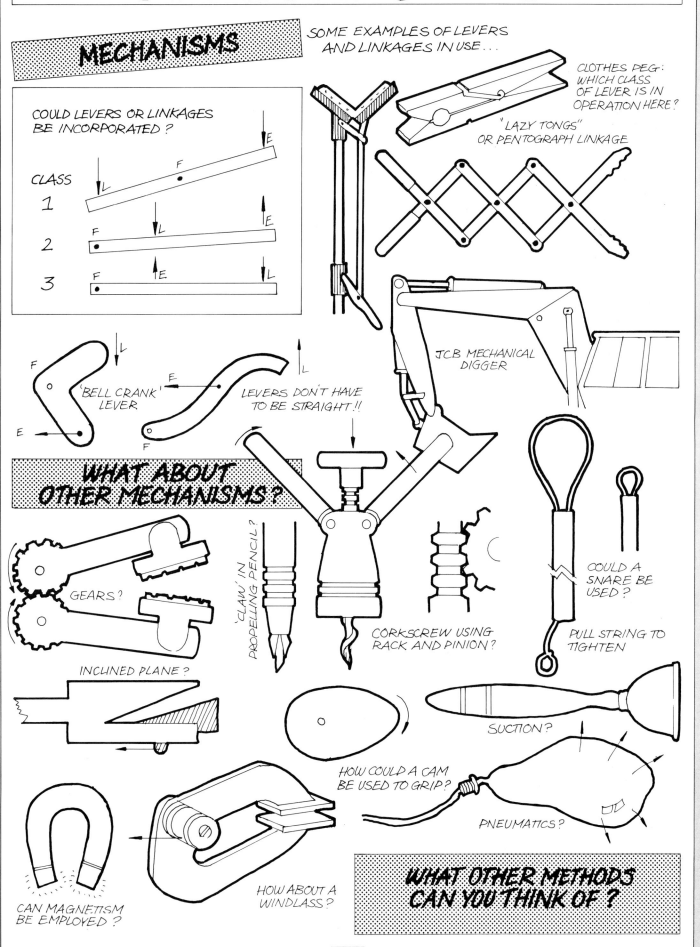

JCB MECHANICAL DIGGER

'BELL CRANK' LEVER

LEVERS DON'T HAVE TO BE STRAIGHT!!

WHAT ABOUT OTHER MECHANISMS?

GEARS?

'CLAW' IN PROPELLING PENCIL?

CORKSCREW USING RACK AND PINION?

COULD A SNARE BE USED?

PULL STRING TO TIGHTEN

INCLINED PLANE?

SUCTION?

HOW COULD A CAM BE USED TO GRIP?

PNEUMATICS?

CAN MAGNETISM BE EMPLOYED?

HOW ABOUT A WINDLASS?

WHAT OTHER METHODS CAN YOU THINK OF?

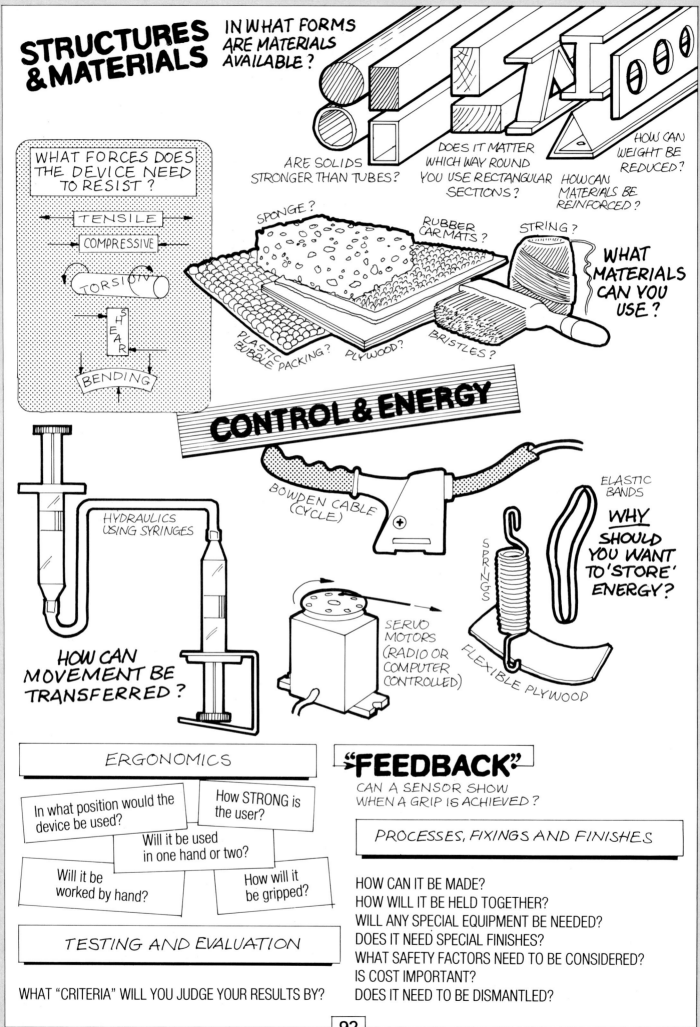

STRUCTURES & MATERIALS

IN WHAT FORMS ARE MATERIALS AVAILABLE?

ARE SOLIDS STRONGER THAN TUBES?

DOES IT MATTER WHICH WAY ROUND YOU USE RECTANGULAR SECTIONS?

HOW CAN MATERIALS BE REINFORCED?

HOW CAN WEIGHT BE REDUCED?

WHAT FORCES DOES THE DEVICE NEED TO RESIST?

- TENSILE
- COMPRESSIVE
- TORSION
- SHEAR
- BENDING

SPONGE?

RUBBER CAR MATS?

STRING?

WHAT MATERIALS CAN YOU USE?

PLASTIC BUBBLE PACKING?

PLYWOOD?

BRISTLES?

CONTROL & ENERGY

HYDRAULICS USING SYRINGES

HOW CAN MOVEMENT BE TRANSFERRED?

BOWDEN CABLE (CYCLE)

SERVO MOTORS (RADIO OR COMPUTER CONTROLLED)

SPRINGS

FLEXIBLE PLYWOOD

ELASTIC BANDS

WHY SHOULD YOU WANT TO 'STORE' ENERGY?

ERGONOMICS

In what position would the device be used?

How STRONG is the user?

Will it be used in one hand or two?

Will it be worked by hand?

How will it be gripped?

"FEEDBACK"

CAN A SENSOR SHOW WHEN A GRIP IS ACHIEVED?

PROCESSES, FIXINGS AND FINISHES

HOW CAN IT BE MADE?
HOW WILL IT BE HELD TOGETHER?
WILL ANY SPECIAL EQUIPMENT BE NEEDED?
DOES IT NEED SPECIAL FINISHES?
WHAT SAFETY FACTORS NEED TO BE CONSIDERED?
IS COST IMPORTANT?
DOES IT NEED TO BE DISMANTLED?

TESTING AND EVALUATION

WHAT "CRITERIA" WILL YOU JUDGE YOUR RESULTS BY?

WHAT IF...?

The way in which we live is constantly changing due to both new technologies and the changing needs of society.

Investigate the way in which a particular technology has developed and suggest ways in which there might be developments in the future. Consider also the side-effects that have occurred in the past and might occur in the future because of this change.

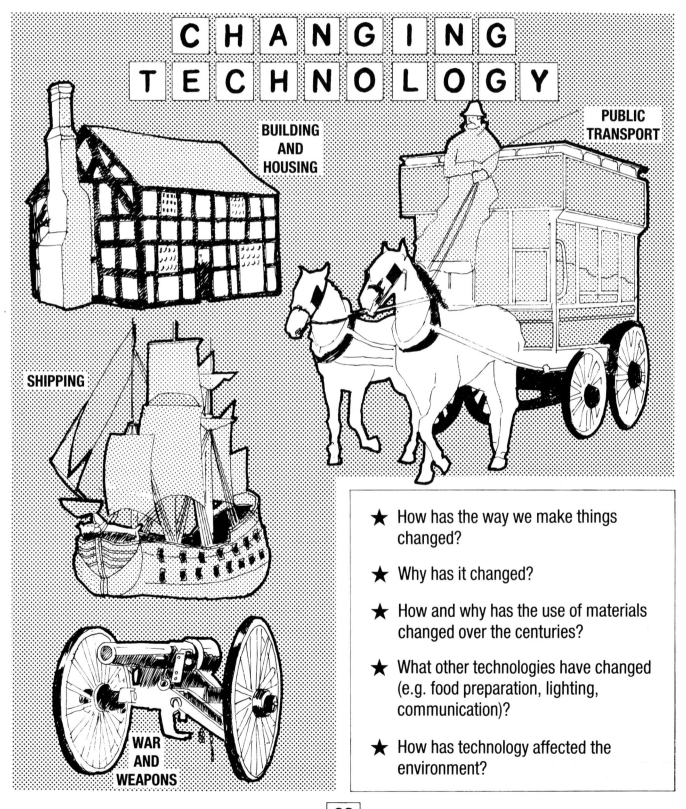

CHANGING TECHNOLOGY

BUILDING AND HOUSING

PUBLIC TRANSPORT

SHIPPING

WAR AND WEAPONS

★ How has the way we make things changed?

★ Why has it changed?

★ How and why has the use of materials changed over the centuries?

★ What other technologies have changed (e.g. food preparation, lighting, communication)?

★ How has technology affected the environment?

C H A N G I N G · S O C I E T Y

Social and technological change often go hand in hand,
creating new problems and needs within society.

What benefits have these
advances in technology
brought to society?

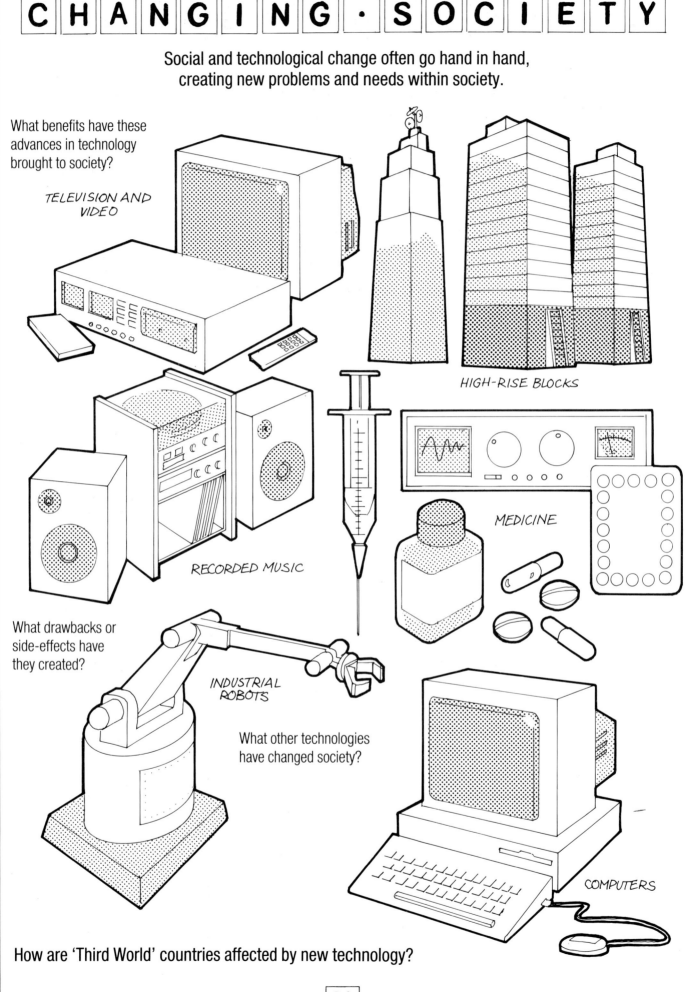

TELEVISION AND
VIDEO

HIGH-RISE BLOCKS

RECORDED MUSIC

MEDICINE

What drawbacks or
side-effects have
they created?

INDUSTRIAL
ROBOTS

What other technologies
have changed society?

COMPUTERS

How are 'Third World' countries affected by new technology?

C·H·A·N·G·I·N·G · S·T·Y·L·E

Consumer products often change purely because of fashion. The style of products usually reflects the popular 'look' of the period, whether this be 'Art Deco', 'Flower Power', or 'High-Tech'. For this reason, old products are being redesigned all the time.

WHY DO PEOPLE SOMETIMES PREFER OLD THINGS TO NEW ONES?

Technology and needs have changed much more rapidly over the last 20 years than ever before.

WHAT MIGHT CHANGE IN THE NEXT 20 YEARS?

COMMUNICATION:

TELEVISION
RADIO
TELEPHONE

FASHION:

STYLE
TEXTILES
SHAPE
COSMETICS

PRODUCT DESIGN:

MINI VIDEO 'CAM-CORDERS'
CAMERAS
MINI COMPUTERS
MOTOR CARS

MUSIC SYSTEMS:

PERSONAL STEREO
'MIDI' HI-FI UNITS
PERSONAL COMPACT DISC
COMPUTER-GENERATED
MUSIC

HOUSE AND HOME:

FURNITURE
KITCHEN EQUIPMENT
FOOD PRESERVATION
LIGHTING
ARCHITECTURE

OR WHAT IF...

WE HAD A
NUCLEAR WAR?
OIL RAN OUT?
WE LIVED IN SPACE?

ACKNOWLEDGEMENTS

The authors would like to thank the following individuals for their permission to use examples of their work:
Page 56: Wen Lan Liang (Bathroom Tidy), Wah Tang (Potentiometers), Jessen Ramesamy (Mechanical Toy); *page 57*: Emily Armer (Mechanical Toy), Wah Tang (Digi-Arm), Donna Mitchell (Electronic Toy), Ruth Williams (Timer); *page 58*: Susan Miles (Buggy Bags); *page 59*: Mark Saunders ("Strida" bicycle); *page 60*: Mark Saunders ("Strida" bicycle), Dominic Rowe (Pen), Carla Junghans (Calculator and Timer), Ruth Williams (Calculator and Torch), Helen Pounds (Stage set and Torch).

We are grateful to the following for permission to reproduce photographs:
Allsport, page 54 *above right* (photo: Tony Duffy); Austin Rover, pages 54 *below left*, 60 *above right*; Dorling Kindersley, London, page 55 *below left and right* from *The Sweater Book*, ed. A. Y. Carroll; Focal Displays Ltd, page 55 *below right*; Harlow Council, Architect's Office, page 54 *centre left*; International Wool Secretariat, page 55 *below left*; Lego UK Ltd, page 60 *above right*; London Borough of Tower Hamlets, page 18 *above left*; Longman Group UK Limited, page 55 *above left and right*.

Longman Group UK Limited,
Longman House, Burnt Mill, Harlow, Essex CM20 2JE, England and associated Companies throughout the world.

First published 1988

ISBN 0 582 02035 2 (ppls)
     0 582 02036 0 (tchrs)

Printed and bound in Great Britain by Butler & Tanner Ltd, Frome and London